THREE PEARLS OF NUMBER THEORY

by

A.Y. Khinchin

Translated by

F. Bagemihl, H. Komm,
and W. Seidel

DOVER PUBLICATIONS, INC.
Mineola, New York

Copyright

Bibliographical Note

This Dover edition, first published in 1998, is an unabridged and unaltered republication of the edition published by The Graylock Press, Baltimore, Maryland in 1952. It was translated from the second, revised Russian edition published in 1948.

Library of Congress Cataloging-in-Publication Data

Khinchin, Aleksandr I͡Akovlevich, 1894–1959.
[Tri zhemchuzhiny teorii chisel. English]
Three pearls of number theory / by A.Y. Khinchin ; translated by F. Bagemihl, H. Komm, and W. Seidel.
p. cm.
"An unabridged and unaltered republication of the edition first published by the Graylock Press, Baltimore, Maryland in 1952. It was translated from the second, revised Russian edition published in 1948"—T.p. verso.
Includes bibliographical references (p. –) and index.
ISBN 0-486-40026-3
1. Number theory. I. Title
QA241.K513 1998
512.'7—dc21 97–48530
CIP

Manufactured in the United States of America
Dover Publications, Inc., 31 East 2nd Street, Mineola, N. Y. 11501

FOREWORD

This little book is devoted to three theorems in arithmetic, which, in spite of their apparent simplicity, have been the objects of the efforts of many important mathematical scholars. The proofs which are presented here make use of completely elementary means, (although they are not very simple).

The book can be understood by beginning college students, and is intended for wide circles of lovers of mathematics.

CONTENTS

A LETTER TO THE FRONT (IN LIEU OF A PREFACE) 9

CHAPTER I. VAN DER WAERDEN'S THEOREM ON ARITHMETIC PROGRESSIONS 11

CHAPTER II. THE LANDAU-SCHNIRELMANN HYPOTHESIS AND MANN'S THEOREM 18

CHAPTER III. AN ELEMENTARY SOLUTION OF WARING'S PROBLEM 37

A LETTER TO THE FRONT

(IN LIEU OF A PREFACE)

March 24, 1945

Dear Seryozha,

Your letter sent from the hospital gave me threefold pleasure. First of all, your request that I send you "some little mathematical pearls" showed that you are really getting well, and are not merely trying, as a brave fighter, to relieve your friends' anxiety. That was my first pleasure.

Furthermore, you gave me occasion to reflect on how it is that in this war such young fighters as you happen to pursue their favorite occupation—the occupation which they cherished already before the war, and from which the war has torn them—so passionately during every little respite. There was nothing like this during the last World War. In those days a young man who had arrived at the ffront almost invariably felt that his life had been disrupted, that what had been the substance of his life before had become for him an unrealizable legend. Now, however, there are some who write dissertations in the intervals between battles, and defend them on their return during a brief furlough! Is it not because you feel with your whole being, that your accomplishments in war and in your favorite occupations—science, art, practical activity—are two links of one and the same great cause? And if so, is not this feeling, perhaps, one of the mainsprings of your victories which we, here at home, are so enthusiastic about? This thought gratified me very much, and that was my second pleasure.

And so I began to think about what to send you. I do not know you very well—you attended my lectures for only one year. Nevertheless I retained a firm conviction of your profound and serious attitude toward science, and therefore I did not want to send you merely some trinkets which were showy but of little substance scientifically. On the other hand I knew that your preparation was not very great—you spent only one year in the university classroom, and during three years of uninterrupted service at the Front you will hardly have had time to study. After several days' deliberation I have made a choice. You must judge for yourself whether it is a happy one or not. Personally I consider the three theorems of arithmetic which I am sending you, to be genuine pearls of our science.

From time to time, remarkable, curious problems turn up in arithmetic, this oldest, but forever youthful, branch of mathematics. In content they are so elementary that any schoolboy can understand them. They are usually concerned with the proof of some very simple law governing the world of numbers, a law which turns out to be correct in all tested special cases. The problem now is to prove that it is in fact always correct. And yet, in spite of the apparent simplicity of the problem, its solution resists, for years and sometimes centuries, the efforts of the most important scholars of the age. You must admit that this is extraordinarily tempting.

I have selected three such problems for you. They have all been solved quite recently, and there are two remarkable common features in their history. First, all three problems have been solved by the most elementary arithmetical methods (do not, however, confuse elementary with simple; as you will see, the solutions of all three problems are not very simple, and it will require not a little effort on your part to understand them well and assimilate them). Secondly, all three problems have been solved by very young, beginning mathematicians, youths of hardly your age, after a series of unsuccessful attacks on the part of "venerable" scholars. Isn't this a spur full of promise for future scholars like you? What an encouraging call to scientific daring!

The work of expounding these theorems compelled me to penetrate more deeply into the structure of their magnificent proofs, and gave me great pleasure.

That was my third pleasure.

I wish you the best of success—in combat and in science.

Yours,

A. Khinchin

CHAPTER I

VAN DER WAERDEN'S THEOREM ON ARITHMETIC PROGRESSIONS

§1

In the summer of 1928 I spent several weeks in Göttingen. As usual, many foreign scholars had arrived there for the summer semester. I got to know many of them, and actually made friends with some. At the time of my arrival, the topic of the day was the brilliant result of a young Hollander, van der Waerden,* who at that time was still a youthful beginner, but is now a well-known scholar. This result had just been obtained here in Göttingen, and in fact, only a few days before my arrival. Nearly all mathematicians whom I met told me about it with enthusiasm.

The problem had the following history. One of the mathematicians there (I forget his name**) had come upon the following problem in the course of his scientific work: Imagine the set of all natural numbers to be divided in any manner whatsoever into two parts (e. g., into even and odd numbers, or into prime and composite numbers, or in any other way). Can one then assert that arithmetic progressions of arbitrary length can be found in at least one of these parts? (By the length of an arithmetic progression I mean here, and in what follows, simply the number of its terms.) All to whom this question was put regarded the problem at first sight as quite simple; its solution in the affirmative appeared to be almost self-evident. The first attempts to solve it, however, led to nought. And as the mathematicians of Göttingen and their foreign guests were by tradition in constant association with one another, this problem, provoking in its resistance, soon became the object of general mathematical interest. Everyone took it up, from the venerable scholar to the young student. After several weeks of strenuous exertions, the problem finally yielded to the attack of a young man who had come to Göttingen to study, the Hollander, van der Waerden. I made his acquaintance, and learned the solution of the problem from him personally. It was elementary, but not simple by any means. The problem turned out to be deep, the appearance of simplicity was deceptive.

*B. L. van der Waerden, *Beweis einer Baudetschen Vermutung*, Nieuw Arch. Wiskunde 15, 212-216 (1927). (Trans.)

**Most probably Baudet; cf. the preceding footnote. (Trans.)

Quite recently, M. A. Lukomskaya (of Minsk) discovered and sent me a considerably simpler and more transparent proof of van der Waerden's theorem, which, with her kind permission, I am going to show you in what follows.

§ 2

Actually, van der Waerden proved somewhat more than what was required. In the first place, he assumes that the natural numbers are divided, not into two, but into arbitrarily many, say k, classes (sets). In the second place, it turns out that it is not necessary to decompose the entire sequence of natural numbers in order to guarantee the existence of an arithmetic progression of prescribed (arbitrarily large) length l in at least one of these classes; a certain segment of it suffices for this purpose. The length, $n(k, l)$, of this segment is a function of the numbers k and l. Of course it doesn't matter where we take this segment, so long as there are $n(k, l)$ successive natural numbers.

Accordingly, van der Waerden's theorem can be formulated as follows:

Let k and l be two arbitrary natural numbers. Then there exists a natural number $n(k, l)$ such that, if an arbitrary segment, of length $n(k, l)$, of the sequence of natural numbers is divided in any manner into k classes (some of which may be empty), then an arithmetic progression of length l appears in at least one of these classes.

This theorem is true trivially for $l=2$. To see this, it suffices to set $n(k, 2)=k+1$; for if $k+1$ numbers are divided into k classes, then certainly at least one of these classes contains more than one number, and an arbitrary pair of numbers forms an arithmetic progression of length 2, which proves the theorem. We shall prove the theorem by induction on l. Consequently, we shall assume throughout the following that the theorem has already been verified for some number $l \geq 2$ and for arbitrary values of k, and shall show that it retains its validity for the number $l+1$ (and naturally also for all values of k).

§ 3

According to our assumption, then, for every natural number k there is a natural number $n(k, l)$ such that, if an arbitrary segment, of length $n(k, l)$, of the natural numbers, is divided in any manner into k classes, there exists in at least one of these classes an arithmetic progression of length l. We must then prove that, for every natural number k, an $n(k, l+1)$ also exists. We solve this problem by actually constructing the number $n(k, l+1)$. To this end we set

$$q_0=1,\ n_0=n(k,l)$$

and then define the numbers $q_1, q_2, \ldots, n_1, n_2, \ldots$ successively as follows: If q_{s-1} and n_{s-1} have already been defined for some $s>0$, we put

$$(1) \qquad q_s=2n_{s-1}q_{s-1}, \quad n_s=n(k^{q_s}, l) \quad (s=1, 2,\ldots).$$

The numbers n_s, q_s are obviously defined hereby for an arbitrary $s\geqq 0$. We now assert that for $n(k, l+1)$ we may take the number q_k. We have to show then that if a segment, of length q_k, of the sequence of natural numbers is divided in any manner into k classes, then there is an arithmetic progression of length $l+1$ in at least one of these classes. The remainder of the chapter is devoted to this proof.

In the sequel we set $l+1=l'$ for brevity.

§ 4

Suppose then that the segment Δ, of length q_k, of the sequence of natural numbers is divided in an arbitrary way into k classes. We say that two numbers a and b of Δ are of the same type, if a and b belong to the same class, and we then write $a\approx b$. Two equally long subsegments of Δ, $\delta=(a, a+1, \ldots, a+r)$ and $\delta'=(a', a'+1, \ldots, a'+r)$, are said to be of the same type, if $a\approx a'$, $a+1\approx a'+1$, ..., $a+r\approx a'+r$, and we then write $\delta\approx\delta'$. The number of different possible types for the numbers of the segment Δ is obviously equal to k. For segments of the form $(a, a+1)$ (i. e., for segments of length 2) the number of possible types is k^2; and in general, for segments of length m, it is k^m. (Of course not all these types need actually appear in the segment Δ.)

Since $q_k=2n_{k-1}q_{k-1}$ (see (1)), the segment Δ can be regarded as a sequence of $2n_{k-1}$ subsegments of length q_{k-1}. Such subsegments, as we have just seen, can have $k^{q_{k-1}}$ different types. The left half of the segment Δ now contains n_{k-1} such subsegments, where $n_{k-1}=n(k^{q_{k-1}}, l)$ according to (1). Because of the meaning of the number $n(k^{q_{k-1}}, l)$, we can assert* that the left half of the segment Δ contains an arithmetic progression of l of these subsegments of

*Work with the initial numbers of the n_{k-1} subsegments. (Trans.)

the same type,

$$\Delta_1, \Delta_2, \ldots, \Delta_l$$

of length q_{k-1}; here we say for brevity that equally long segments Δ_i form an arithmetic progression, if their initial numbers form such a progression. We call the difference between the initial numbers of two neighboring segments of the progression $\Delta_1, \Delta_2, \ldots, \Delta_l$ the difference d_1 of this progression. Naturally the difference between the second (or third, fourth, etc.) numbers of two such neighboring segments is likewise equal to d_1.

To this progression of segments we now add the succeeding, $(l+1)$-st, term $\Delta_{l'}$ (we recall that $l'=l+1$) which may already project beyond the boundary of the left half of the segment Δ, but which in any case still belongs entirely to the segment Δ. The segments $\Delta_1, \Delta_2, \ldots, \Delta_l, \Delta_{l'}$ then form an arithmetic progression, of length $l'=l+1$ and difference d_1, of segments of length q_{k-1}, where $\Delta_1, \Delta_2, \ldots, \Delta_l$ are of the same type. We know nothing about the type of the last segment $\Delta_{l'}$.

This completes the first step of our construction. It would be well if you thought it through once more before we continued.

§5

We now proceed to the second step. We take an arbitrary one of the first l terms of the progression of segments just constructed. Let this term be Δ_{i_1}, so that $1 \leqq i_1 \leqq l$; Δ_{i_1} is a segment of length q_{k-1}. We treat it the same way as we treated the segment Δ. Since $q_{k-1} = 2n_{k-2}q_{k-2}$, the left half of the segment Δ_{i_1} can be regarded as a sequence of n_{k-2} subsegments of length q_{k-2}. For subsegments of this length there are $k^{q_{k-2}}$ types possible, and on the other hand $n_{k-2} = n(k^{q_{k-2}}, l)$ because of (1). Therefore the left half of Δ_{i_1} must contain a progression of l of these subsegments of the same type, $\Delta_{i_1 i_2}$ $(1 \leqq i_2 \leqq l)$, of length q_{k-2}. Let d_2 be the difference of this progression (i.e., the distance between the initial numbers of two neighboring segments). To this progression of segments we add the $(l+1)$-st term $\Delta_{i_1 l'}$, about whose type, of course, we know nothing. The segment $\Delta_{i_1 l'}$ does not have to belong to the left half of the segment Δ_{i_1} any more, but must obviously belong to the segment Δ_{i_1}.

We now carry over our construction, which we have executed up to now in only one of the segments Δ_{i_1}, congruently to all the other segments Δ_{i_1} $(1 \leqq i_1 \leqq l')$. We thus obtain a set of segments $\Delta_{i_1 i_2}$ $(1 \leqq i_1 \leqq l',\ 1 \leqq i_2 \leqq l')$ with two indices. It is clear that two arbitrary segments of this set with indices not exceeding l are of the same type:

$$\Delta_{i_1 i_2} \approx \Delta_{i_1' i_2'} \qquad (1 \leqq i_1,\ i_2,\ i_1',\ i_2' \leqq l)$$

You no doubt see now that this process can be continued. We carry it out k times. The results of our construction after the first step were segments of length q_{k-1}, after the second step, segments of length q_{k-2}, etc. After the k-th step, therefore, the results of the construction are segments of length $q_0 = 1$, i. e., simply numbers of our original segment Δ. Nevertheless we denote them as before by

$$\Delta_{i_1 i_2 \ldots i_k} \quad (1 \leqq i_1,\ i_2,\ \ldots,\ i_k \leqq l').$$

For $1 \leqq s \leqq k$ and $1 \leqq i_1, \ldots, i_s, i_1', \ldots, i_s' \leqq l$ we have

(2) $$\Delta_{i_1 i_2 \ldots i_s} \approx \Delta_{i_1' i_2' \ldots i_s'}$$

We now make two remarks which are important for what follows.

1) In (2), if $s < k$ and if $i_{s+1}, i_{s+2}, \ldots, i_k$ are arbitrary indices taken from the sequence 1, 2, ..., l, l', then the number $\Delta_{i_1 i_2 \ldots i_s i_{s+1} \ldots i_k}$ appears in the same position in the segment $\Delta_{i_1 \ldots i_s}$ as the number $\Delta_{i_1' i_2' \ldots i_s' i_{s+1} \ldots i_k}$ does in the segment $\Delta_{i_1' \ldots i_s'}$. Since these two segments are of the same type because of (2), it follows that

(3) $$\Delta_{i_1 i_2 \ldots i_s i_{s+1} \ldots i_k} \approx \Delta_{i_1' i_2' \ldots i_s' i_{s+1} \ldots i_k}$$

if $1 \leqq i_1, \ldots, i_s, i_1', \ldots, i_s' \leqq l$ and $1 \leqq i_{s+1}, i_{s+2}, \ldots, i_k \leqq l'$ $(1 \leqq s \leqq k)$.

2) For $s \leqq k$ and $i_s' = i_s + 1$, $\Delta_{i_1 \ldots i_{s-1} i_s}$ and $\Delta_{i_1 \ldots i_{s-1} i_s'}$ are obviously neighboring segments in the s-th step of our construction. Therefore for arbitrary indices $i_{s+1}, \ldots, i_k$, the numbers $\Delta_{i_1 \ldots i_{s-1} i_s i_{s+1} \ldots i_k}$ and $\Delta_{i_1 \ldots i_{s-1} i_s' i_{s+1} \ldots i_k}$ appear in the same position in two such neighboring segments, so that (with $i_s' = i_{s+1}$)

(4) $$\Delta_{i_1 \ldots i_{s-1}\, i'_s\, i_{s+1} \ldots i_k} - \Delta_{i_1 \ldots i_{s-1}\, i_s\, i_{s+1} \ldots i_k} = d_s.$$

§6

Now we are near our goal. We consider the following $k+1$ numbers of the segment Δ:

(5) $$\begin{cases} a_0 = \Delta_{l' l' l' \ldots l'} \\ a_1 = \Delta_{1\, l' l' \ldots l'} \\ a_2 = \Delta_{1\, 1\, l' \ldots l'} \\ \cdot\ \cdot\ \cdot\ \cdot\ \cdot\ \cdot\ \cdot\ \cdot \\ a_k = \Delta_{1\, 1\, 1 \cdots 1}\ . \end{cases}$$

Since the segment Δ has been divided into k classes, and we have $k+1$ numbers in (5), there are two of these numbers which belong to the same class. Let these be the numbers a_r and a_s $(r<s)$, so that

(6) $$\Delta_{\underbrace{1 \ldots 1}_{r}\, \underbrace{l' \ldots l'}_{k-r}} \approx \Delta_{\underbrace{1 \ldots 1}_{s}\, \underbrace{l' \ldots l'}_{k-s}}\ .$$

We consider the $l+1$ numbers

(7) $$c_i = \Delta_{\underbrace{1 \ldots 1}_{r}\, \underbrace{i \ldots i}_{s-r}\, \underbrace{l' \ldots l'}_{k-s}} \qquad (1 \leqq i \leqq l')\ :$$

The first l numbers of this group (i. e., those with $i<l'$) belong to the same class because of (3). The last $(i=l')$, however, is of the same type as the first because of (6). Consequently all $l+1$ numbers in (7) are of the same type, and to prove our assertion we have only to show that these numbers form an arithmetic progression, i. e., that the difference $c_{i+1} - c_i$ $(1 \leqq i \leqq l)$ does not depend on i.

We set $i+1 = i'$ for brevity. Further let

$$c_{i,m} = \Delta_{\underbrace{1 \ldots 1}_{r}\, \underbrace{i' \ldots i'}_{m}\, \underbrace{i \ldots i}_{s-r-m}\, \underbrace{l' \ldots l'}_{k-s}} \qquad (0 \leqq m \leqq s-r),$$

so that $c_{i,0} = c_i$, $c_{i,s-r} = c_{i+1}$, and hence

$$c_{i+1} - c_i = \sum_{m=1}^{s-r} (c_{i,m} - c_{i,m-1}).$$

Because of (4) we have

$$c_{i,m} - c_{i,m-1} = \Delta_{\underbrace{1\ldots1}_{r}\,\underbrace{i'\ldots i'}_{m}\,\underbrace{i\ldots i}_{s-r-m}\,\underbrace{l'\ldots l'}_{k-s}} - \Delta_{\underbrace{1\ldots1}_{r}\,\underbrace{i'\ldots i'}_{m-1}\,\underbrace{i\ldots i}_{s-r-m+1}\,\underbrace{l'\ldots l'}_{k-s}} = d_{r+m}.$$

Thus the difference

$$c_{i+1} - c_i = d_{r+1} + d_{r+2} + \ldots + d_s,$$

and is indeed independent of i, which completes the proof of our assertion.

You see how complicated a completely elementary construction can sometimes be. And yet this is not an extreme case: in the next chapter you will encounter just as elementary a construction which is considerably more complicated. Besides, it is not out of the question that van der Waerden's theorem admits of an even simpler proof, and all research in this direction can only be welcomed.

CHAPTER II

THE LANDAU-SCHNIRELMANN HYPOTHESIS AND MANN'S THEOREM

§1

You have perhaps heard of the remarkable theorem of Lagrange, that *every natural number is the sum of at most four squares*. In other words, every natural number is either itself the square of another number, or else the sum of two, or else of three, or else of four such squares. For the purpose at hand it is desirable to understand the content of this theorem in a somewhat different form. Let us write down the sequence of all perfect squares, beginning with zero:

$$(S) \qquad 0, 1, 4, 9, 16, 25, \dots .$$

This is a certain sequence of whole numbers. We denote it by S, and imagine four completely identical copies of it, S_1, S_2, S_3, S_4, to be written down. Now we choose an arbitrary number a_1^2 from S_1, an arbitrary number a_2^2 from S_2, an arbitrary number a_3^2 from S_3 and an arbitrary number a_4^2 from S_4, and add these numbers together. The resulting sum

$$(*) \qquad n = a_1^2 + a_2^2 + a_3^2 + a_4^2$$

can be

1) zero (if $a_1 = a_2 = a_3 = a_4 = 0$);

2) the square of a natural number (if in some representation (*) of the number n three of the numbers a_1, a_2, a_3, a_4 are zero and the fourth is not zero);

3) the sum of two squares of natural numbers (if in some representation (*) of the number n two of the numbers a_1, a_2, a_3, a_4 are zero and the other two are not zero);

4) the sum of three squares of natural numbers (if in some representation (*) of the number n one of the numbers a_1, a_2, a_3, a_4 is equal to zero and the remaining three are not zero);

5) the sum of four squares of natural numbers (if in some representation of the number n all four numbers are different from zero).

Thus the resulting number n is either zero or else a natural number which can be represented as the sum of at most four squares, and it is clear that conversely every natural number can be obtained by the process which we have described.

Now let us arrange all natural numbers n which can be obtained by means of our process (i.e., by the addition of four numbers taken respectively from the sequences S_1, S_2, S_3, S_4), in order of magnitude, in the sequence

$$(A) \qquad 0, n_1, n_2, n_3, \dots$$

(where $0<n_1<n_2<n_3<\dots$, so that if there are equal numbers among those constructed, only one of them appears in (A)). The theorem of Lagrange now asserts simply that the sequence (A) contains all the natural numbers, i. e., that $n_1=1$, $n_2=2$, $n_3=3$, etc.

We shall now generalize our process. Let there be given k monotonically increasing sequences of integers which begin with zero:

$$(A^{(1)}) \qquad 0, a_1^{(1)}, a_2^{(1)}, \dots, a_m^{(1)}, \dots,$$

$$(A^{(2)}) \qquad 0, a_1^{(2)}, a_2^{(2)}, \dots, a_m^{(2)}, \dots,$$

$$\dots\dots \qquad . \; . \; . \; . \; . \; . \; . \; . \; . \; . \; . \; .$$

$$(A^{(k)}) \qquad 0, a_1^{(k)}, a_2^{(k)}, \dots, a_m^{(k)}, \dots .$$

We choose arbitrarily a single number from each sequence $A^{(i)}$ $(1 \leqq i \leqq k)$ and add these k numbers together. The totality of all numbers constructed in this manner, if we order them according to magnitude, yields a new sequence

$$(A) \qquad 0, n_1, n_2, \dots, n_m, \dots$$

of the same type, which we shall call the *sum* of the given sequences $A^{(1)}$, $A^{(2)}$, ..., $A^{(k)}$:

$$A=A^{(1)}+A^{(2)}+\dots+A^{(k)}=\sum_{i=1}^{k} A^{(i)} .$$

The content of Lagrange's theorem is that the sum $S+S+S+S$ contains the entire sequence of natural numbers.

Perhaps you have heard of the famous theorem of Fermat, that *the sum* $S+S$ *contains all prime numbers which leave a remainder of* 1 *when divided by* 4 (i. e., the numbers 5, 13, 17, 29, ...). Perhaps you also know that the famous Soviet scholar Ivan Matveyevitch Vinogradov proved the following remarkable theorem, on which many of the greatest mathematicians of the preceding two centuries had worked without success:

If we denote by P *the sequence*

$$(P) \qquad 0, 2, 3, 5, 7, 11, 13, 17, \ldots$$

consisting of zero and all prime numbers, then the sum $P+P+P$ *contains all sufficiently large odd numbers.*

I have cited all these examples here for only one very modest purpose: to familiarize you with the concept of the sum of sequences of numbers and to show how some classical theorems of number theory can be formulated simply and conveniently with the aid of this concept.

§2

As you have undoubtedly observed, in all the examples mentioned we are concerned with showing that the sum of a certain number of sequences represents a sequence which contains either completely or almost completely this or that class of numbers (e. g., all the natural numbers, all sufficiently large odd numbers, and others of the same sort). In all other similar problems the purpose of the investigation is to prove that the sum of the given sequences of numbers represents a set of numbers which is in some sense "dense" in the sequence of natural numbers. It is often the case that this set contains the entire sequence of natural numbers (as we saw in our first example). The theorem of Lagrange says that the sum of the four sequences S contains the whole sequence of natural numbers. Now it is customary to call the sequence A a *basis* (of the sequence of natural numbers) *of order* k if the sum of k identical sequences A contains all the natural numbers. The theorem of Lagrange then states that the sequence S of perfect squares is a basis of order four. It was shown later that the sequence of perfect cubes forms a basis of order nine. A little reflection shows that every basis of order k is also a basis of order $k+1$.

In all these and in many other examples the "density" of the sum which is to be established is determined by particular properties of the sequences that are added together, i. e., by the special arithmetical nature of the numbers which go to make up these sequences (these numbers being either perfect squares, or primes, or others of a similar nature). Sixteen years ago the distinguished Soviet scholar Lev Genrichovitch Schnirelmann first raised the question: To what extent is the density of the sum of several sequences determined solely by the density of the summands, irrespective of their arithmetical nature. This problem turned out to be not only deep and interesting, but also useful for the treatment of some classical problems. During the intervening fifteen years it received the attention of many outstanding scholars, and it has given rise to a rich literature.

Before we can state problems in this field precisely and write the word "density" without quotation marks, it is evident that we must first agree on what number (or on what numbers) to use to measure the "density" of our sequences with (just as in physics the words "warm" and "cold" do not acquire a precise scientific meaning until we have learned to measure temperature).

A very convenient measure of the "density" of a sequence of numbers, which is now used for all scientific problems of the kind we are considering, was proposed by L. G. Schnirelmann. Let

$$(A) \qquad 0, a_1, a_2, \ldots, a_n, \ldots$$

be a sequence of numbers, where, as usual, all the a_n are natural numbers and $a_n < a_{n+1}$ $(n = 1, 2, \ldots)$. We denote by $A(n)$ the number of natural numbers in the sequence (A) which do not exceed n (zero is not counted), so that $0 \leqq A(n) \leqq n$. Then the inequality

$$0 \leqq \frac{A(n)}{n} \leqq 1$$

holds. The fraction $A(n)/n$, which for different n naturally has different values, can obviously be interpreted as a kind of average density of the sequence (A) in the segment from 1 to n of the sequence of natural numbers. Following the suggestion of Schnirelmann, the *greatest lower bound* of all values of this fraction is called the *density* of the sequence (A) (in the entire sequence of natural numbers). We shall denote this density by $d(A)$.

In order to become familiar at once with the elementary properties of this concept, I recommend that you convince yourself of the validity of the following theorems:

1. If $a_1 > 1$ (i. e., the sequence (A) does not contain unity), then $d(A) = 0$.

2. If $a_n = 1 + r(n-1)$ (i. e., the sequence (A), beginning with a_1, is an arithmetic progression with initial term 1 and difference r), then $d(A) = 1/r$.

3. The density of every geometric progression is equal to zero.

4. The density of the sequence of perfect squares is equal to zero.

5. For the sequence (A) to contain the entire sequence of natural numbers ($a_n = n$, $n = 1, 2, \ldots$), it is necessary and sufficient that $d(A) = 1$.

6. If $d(A) = 0$ and A contains the number 1, and if $\epsilon > 0$ is arbitrary, then there exists a sufficiently large number m such that $A(m) < \epsilon m$.

If you have proved all this, you are familiar enough with the concept of density to be able to use it. Now I want to acquaint you with the proof of the following remarkable, albeit very simple, lemma of Schnirelmann:

$$d(A+B) \geqq d(A) + d(B) - d(A)\, d(B). \tag{1}$$

The meaning of this inequality is clear: the density of the sum of two arbitrary sequences of numbers is not smaller than the sum of their densities diminished by the product of these densities. This "Schnirelmann inequality" represents the first tool, still crude to be sure, for estimating the density of a sum from the densities of the summands. Here is its proof. We denote by $A(n)$ the number of natural numbers which appear in the sequence A and do not exceed n, and by $B(n)$ the analogous number for the sequence B. For brevity we set $d(A) = \alpha$, $d(B) = \beta$, $A + B = C$, $d(C) = \gamma$. The segment $(1, n)$ of the sequence of natural numbers contains $A(n)$ numbers of the sequence A, each of which also appears in the sequence C. Let a_k and a_{k+1} be two consecutive numbers of this group. Between them there are $a_{k+1}-a_k-1=l$ numbers which do not belong to A. These are the numbers

$$a_k+1,\ a_k+2,\ \ldots,\ a_k+l=a_{k+1}-1.$$

Some of them appear in C, e.g., all numbers of the form a_k+r, where r occurs in B (which we abbreviate as follows: $r \varepsilon B$). There are as many numbers of this last kind, however, as there are numbers of B in the segment $(1, l)$, that is, $B(l)$ of them. Consequently every segment of length l included between two consecutive numbers of the sequence A contains at least $B(l)$ numbers which belong to C. It follows that the number, $C(n)$, of numbers of the segment $(1, n)$ appearing in C is at least

$$A(n)+\Sigma\, B(l)$$

where the summation is extended over all segments which are free of the numbers appearing in A. According to the definition of density, however, $B(l) \geqq \beta l$, so that

$$C(n) \geqq A(n)+\beta\Sigma l = A(n)+\beta\{n-A(n)\},$$

because Σl is the sum of the lengths of all the segments which are free of the numbers appearing in A, which is simply the number $n-A(n)$ of numbers of the segment $(1, n)$ which do not occur in A. But $A(n) \geqq \alpha n$, and hence

$$C(n) \geqq A(n)(1-\beta)+\beta n \geqq \alpha n(1-\beta)+\beta n,$$

which yields

$$C(n)/n \geqq \alpha+\beta-\alpha\beta.$$

Since this inequality holds for an arbitrary natural number n, we have

$$\gamma = d(C) \geqq \alpha+\beta-\alpha\beta, \qquad \text{Q. E. D.}$$

Schnirelmann's inequality (1) can be written in the equivalent form

$$1-d(A+B)\leqq\{1-d(A)\}\{1-d(B)\},$$

and in this form can easily be generalized to the case of an arbitrary number of summands:

$$1-d(A_1+A_2+\ldots+A_k)\leqq\prod_{i=1}^{k}\{1-d(A_i)\}.$$

It is proved by a simple induction; you should have no trouble in carrying it out yourself. If we write the last inequality in the form

$$d(A_1+A_2+\ldots+A_k)\geqq 1-\prod_{i=1}^{k}\{1-d(A_i)\}, \tag{2}$$

it again enables one to estimate the density of a sum from the densities of the summands. L. G. Schnirelmann derived a series of very remarkable results from his elementary inequality, and obtained above all the following important theorem:

Every sequence of positive density is a basis of the sequence of natural numbers.

In other words, if $\alpha=d(A)>0$, then the sum of a sufficiently large number of sequences A contains the entire sequence of natural numbers. The proof of this theorem is so simple that I should like to tell you about it, even though this will divert us a bit from our immediate problem.

Let us denote for brevity by A_k the sum of k sequences, each of which coincides with A. Then by virtue of inequality (2),

$$d(A_k)\geqq 1-(1-\alpha)^k.$$

Since $\alpha>0$, we have, for sufficiently large k,

$$d(A_k)>\tfrac{1}{2}. \tag{3}$$

Now one can easily show that the sequence A_{2k} contains the whole sequence of natural numbers. This is a simple consequence of the following general proposition.

LEMMA. *If* $A(n)+B(n)>n-1$, *then* n *occurs in* $A+B$.

Indeed, if n appears in A or in B, everything is proved. We may therefore assume that n occurs in neither A nor B. Then $A(n) = A(n-1)$ and $B(n) = B(n-1)$, and consequently

$$A(n-1)+B(n-1)>n-1.$$

Now let $a_1, a_2, \ldots, a_r$ and $b_1, b_2, \ldots, b_s$ be the numbers of the segment $(1, n-1)$ which appear in A and B, respectively, so that $r = A(n-1)$, $s = B(n-1)$. Then all the numbers

$$a_1, a_2, \ldots, a_r,$$
$$n-b_1, n-b_2, \ldots, n-b_s$$

belong to the segment $(1, n-1)$. There are $r+s = A(n-1)+B(n-1)$ of these numbers, which is more than $n-1$. Hence one of the numbers in the upper row equals one of the numbers in the lower row. Let $a_i = n - b_k$. Then $n = a_i + b_k$, i. e., n appears in $A+B$.

Returning now to our objective, we have, on the basis of (3), for an arbitrary n:

$$A_k(n) > \tfrac{1}{2}n > \tfrac{1}{2}(n-1)$$

and therefore

$$A_k(n)+A_k(n)>n-1.$$

According to the lemma just proved, it follows that n appears in $A_k + A_k = A_{2k}$. But n is an arbitrary natural number, and hence our theorem is proved.

This simple theorem led to a series of important applications in the papers of L. G. Schnirelmann. For example, he was the first to prove that the sequence P consisting of unity and all the prime numbers is a basis of the sequence of natural numbers. The sequence P, it is true, has density zero, as Euler had already shown, so that the theorem which we just proved is not directly applicable to it. But Schnirelmann was able to prove that $P+P$ has positive density. Hence $P+P$ forms a basis, and therefore P indeed also. From this it is easy to infer that an arbitrary natural number, with the exception of 1, can, for sufficiently large k, be represented as the sum of at most k primes. For that time (1930) this result was fundamental and evoked the greatest interest in the scientific world. At present, thanks to the remarkable work of I. M. Vinogradov, we know considerably more in this direction, as I already related to you at the beginning of this chapter.

§3

In the preceding it was my purpose to introduce you in the shortest way possible to the problems of this singular and fascinating branch of number theory, whose study began with L. G. Schnirelmann's remarkable work. The immediate goal of the present chapter, however, is a specific problem in this field, and I now proceed to its formulation.

In the fall of 1931, upon his return from a foreign tour, L. G. Schnirelmann reported to us his conversations with Landau in Göttingen, and related among other things that in the course of these conversations they had discovered the following interesting fact: In all the concrete examples that they were able to devise, it was possible to replace the inequality

$$d(A+B) \geqq d(A)+d(B)-d(A)\,d(B),$$

which we derived in §2, by the sharper (and simpler) inequality

$$d(A+B) \geqq d(A)+d(B). \tag{4}$$

That is, the density of the sum always turned out to be at least as large as the sum of the densities of the summands (under the assumption, of course, that $d(A)+d(B) \leqq 1$). They therefore naturally assumed that inequality (4) was the expression of a universal law, but the first attempts to prove this conjecture were unsuccessful. It soon became evident that if their conjecture was correct, the road to its proof would be quite difficult. We wish to note at this point that if the hypothetical inequality (4) does represent a universal law, then this law can be generalized immediately by induction to the case of an arbitrary number of summands; i.e., under the assumption that

$$\sum_{i=1}^{k} d(A_i) \leqq 1$$

we have

$$d\Big(\sum_{i=1}^{k} A_i\Big) \geqq \sum_{i=1}^{k} d(A_i). \tag{5}$$

This problem could not help but attract the attention of scholars, because of the simplicity and elegance of the general hypothetical law (4) on the one hand, and on the other because of the sharp contrast between the elementary character of the problem and the difficulty of its solution which became apparent already after the first attacks. I myself was fascinated by it at the time, and neglected all my other researches on its account. Early in 1932, after several months of hard work, I succeeded in proving inequality (4) for the most important special case, $d(A)=d(B)$ (this case must be considered as the most important because in the majority of concrete problems all the summands are the same). At the same time I also proved the general inequality (5) under the assumption that $d(A_1)=d(A_2)=\ldots=d(A_k)$ (it is easy to see that this result cannot be derived from the preceding one simply by induction, but requires a special proof). The method which I used was completely elementary, but very complicated. I was later able to simplify the proof somewhat.

Be that as it may, it was but a special case. For a long time it seemed to me that a none too subtle improvement of my method should lead to a full solution of the problem, but all my efforts in this direction proved fruitless.

In the meantime the publication of my work had attracted the attention of a wide circle of scholars in all countries to the Landau-Schnirelmann hypothesis. Many insignificant results were obtained, and a whole literature sprang up. Some authors carried over the problem from the domain of natural numbers to other fields. In short, the problem became "fashionable". Learned societies offered prizes for its solution. My friends in England wrote me in 1935 that a good half of the English mathematicians had postponed their usual work in order to try to solve this problem. Landau, in his tract devoted to the latest advances in additive number theory, wrote that he "should like to urge this problem on the reader". But it proved to be obstinate, and withstood the efforts of the most able scholars for a whole series of years. It was not until 1942 that the young American mathematician Mann finally disposed of it: he found a complete proof of inequality (4) (and hence also of inequality (5)). His method is wholly elementary and is related to my work in form, although it is based on an entirely different idea. The proof is long and very complicated, and I could not bring myself to present it to you here. A year later, however, in 1943, Artin and Scherk published a new proof of the same theorem, which rests on an altogether different idea. It is considerably shorter and more transparent, though still quite elementary. This is the proof that I should like to tell you about; I have written this chapter on its account, and it forms the content of all the succeeding sections.

§4

Suppose then that A and B are two sequences. We set $A+B=C$. Let $A(n)$, $d(A)$, etc. have their usual meaning. We recall that all our sequences begin with zero, but that only the natural numbers appearing in these sequences are considered when calculating $A(n)$, $B(n)$, $C(n)$. We have to prove that

$$d(C)\geqq d(A)+d(B) \tag{6}$$

provided that $d(A)+d(B)\leqq 1$. For brevity we set $d(A)=\alpha$, $d(B)=\beta$ in what follows.

Fundamental Lemma. *If n is an arbitrary natural number, there exists an integer m $(1\leqq m\leqq n)$ such that*

$$C(n)-C(n-m)\geqq(\alpha+\beta)\,m.$$

In other words, there exists a "remainder" $(n-m+1, n)$ of the segment $(1, n)$, in which the average density of the sequence C is at least $\alpha+\beta$.

We are now faced with two problems: first, to prove the fundamental lemma, and second, to show that inequality (6) follows from the fundamental lemma. The second of these problems is incomparably easier than the first, and we shall therefore begin with the second problem.

Suppose then that the fundamental lemma has already been proved. This means that in a certain "remainder" $(n-m+1, n)$ of the segment $(1, n)$ the average density of the sequence C is at least $\alpha+\beta$. By the fundamental lemma, however, the segment $(1, n-m)$ again has a certain "remainder" $(n-m-m'+1, n-m)$ in which the average density of C is at least $\alpha+\beta$. It is clear that by continuing this process, the segment $(1, n)$ is eventually divided into a finite number of subsegments, in each of which the average density of C is at least $\alpha+\beta$. Therefore the average density of C is also at least $\alpha+\beta$ in the whole segment $(1, n)$. Since n was arbitrary, however, we have

$$d(C) \geqq \alpha+\beta, \qquad \text{Q.E.D.}$$

Thus the problem is now reduced to proving the fundamental lemma. We now turn to this proof, which is long and complicated.

§5
NORMAL SEQUENCES

In all that follows we shall regard the number n as fixed, and all sequences which we investigate will consist of the number 0 and certain numbers of the segment $(1, n)$. We agree to call such a sequence N *normal*, if it possesses the following property: If the arbitrary numbers f and f' of the segment $(1, n)$ do not appear in N, then neither does the number $f+f'-n$ appear in N (where the case $f=f'$ is not excluded).

If the number n belongs to the sequence C, then

$$C(n)-C(n-1)=1 \geqq (\alpha+\beta)\cdot 1,$$

so that the fundamental lemma is trivially correct ($m=1$). Consequently we shall assume in the sequel—I beg you to keep this in mind—that n does not occur in C.

To begin with, the fundamental lemma is easy to prove in case the sequence C is normal. Indeed, let us denote by m the smallest positive number which does not appear in C ($m \leqq n$ because n, by assumption, does not occur in C). Let s be an arbitrary integer lying between $n-m$ and n; $n-m<s<n$. Then $0<s+m-n<m$. I say that $s \varepsilon C$. For if this were not the case, the number $s+m-n$, because of the normality of C, would not appear in C. But we have just seen that this number is smaller than m, whereas m, by definition, is the *smallest* positive integer which does not occur in C.

Hence all numbers s of the segment $n-m<s<n$ appear in C, and therefore

$$C(n)-C(n-m)=m-1.$$

On the other hand, by the lemma on p. 24, since m does not occur in $C=A+B$ we have $A(m)+B(m)\leqq m-1$. Consequently

$$C(n)-C(n-m)\geqq A(m)+B(m)\geqq(\alpha+\beta)m, \tag{7}$$

which again proves the validity of the fundamental lemma.

§6
CANONICAL EXTENSIONS

We now turn our attention to the case where the sequence $C=A+B$ is not normal. In this case we shall add to the set B, according to a very definite scheme, numbers which it does not contain, and thereby pass from B to an extended set B_1. The set $A+B_1=C_1$ evidently will then be a certain extension of the set C. As I said before, this extension of the sets B and C (the set A remains unaltered) will be defined precisely and unambiguously; it is possible if and only if the set C is not normal. We shall call this extension a *canonical extension* of the sets B and C. Some important properties of canonical extensions will be derived, with whose aid the proof of the fundamental lemma will be completed.

I now come to the definition of the canonical extension of the sets B and C. If C is not normal, there exist two numbers c and c' in the segment $(0,n)$, such that

$$c\notin C,\quad c'\notin C,\quad c+c'-n\in C.$$

Since $C=A+B$, it follows that

$$c+c'-n=a+b \qquad (a\in A,\ b\in B). \tag{8}$$

Let β_0 be the smallest number of the set B which can play the role of the number b in equation (8). In other words, β_0 is the smallest integer $b\in B$ which satisfies equation (8) for suitably chosen numbers $c\notin C$, $c'\notin C$, $a\in A$ of the segment $(0,n)$. This number β_0 will be called the *basis* of our extension.

Thus the equation

$$c+c'-n=a+\beta_0 \tag{9}$$

necessarily has solutions in the numbers c, c', a satisfying the conditions

$$c \notin C, \; c' \notin C, \; a \in A,$$

where all three numbers belong to the segment $(0, n)$. We write all numbers c and c' which satisfy equation (9) and the enumerated conditions, to form a set C^*. Clearly the sets C and C^* do not have a single element in common. We call their union* (i. e., the totality of all numbers which occur either in C or in C^*)

$$C \cup C^* = C_1$$

the *canonical extension* of the set C.

Let us now examine the expression $\beta_0 + n - c$ If c here is allowed to run through all the numbers of the set C^* just constructed, the values of this expression form a certain set B^*. According to equation (9), every such number $\beta_0 + n - c$ $(c \in C^*)$ can be written in the form $c' - a$, where $c' \in C^*$, $a \in A$.

Let b^* be an arbitrary number occurring in B^*. Since it is of the form $\beta_0 + n - c$, it is $\geqq \beta_0 \geqq 0$; and since it is also of the form $c' - a$ $(c' \in C^*, \; a \in A)$, it is $\leqq c' \leqq n$. Hence all numbers of the set B^* belong to the segment $(0, n)$. Moreover, if $b^* \in B^*$, then $b^* \notin B$, because otherwise it would follow from $b^* = c' - a$ that $c' = a + b^* \in A + B = C$, which is false.

Accordingly, the set B^* is embedded in the segment $(0, n)$ and has no elements in common with the set B. We put

$$B \cup B^* = B_1$$

and call the set B_1 a canonical extension of the set B.

Let us show that

$$A + B_1 = C_1.$$

First, let $a \in A$, $b_1 \in B_1$. We shall prove that $a + b_1 \in C_1$. From $b_1 \in B_1$ it follows that either $b_1 \in B$ or $b_1 \in B^*$. If $b_1 \in B$, then $a + b_1 \in A + B = C \subset C_1$. If $b_1 \in B^*$, however, then $a + b_1$ either occurs in C, and hence also in C_1, or $a + b_1 \notin C$. But in this case (since b_1, as an element of the set B^*, is of the form $\beta_0 + n - c'$, $c' \notin C$) we obtain

$$c = a + b_1 = a + \beta_0 + n - c' \notin C.$$

Therefore

*Here and in the sequel we use the symbol $\cup$ to denote the union of sets, since we are using the symbol + in another sense.

$$c+c'-n=a+\beta_0 \varepsilon A+B=C,$$

where $c \notin C$ and $c' \notin C$. But then according to the definition of the set C^*,

$$c=a+b_1 \varepsilon C^* \subset C_1, \qquad \text{Q.E.D.}$$

Thus we have shown that $A+B_1 \subset C_1$.

To prove the inverse relation, let us assume that $c \varepsilon C_1$, which means that either $c \varepsilon C$ or $c \varepsilon C^*$. If $c \varepsilon C$, then $c=a+b$, $a \varepsilon A$, $b \varepsilon B \subset B_1$. If, however, $c \varepsilon C^*$, then, for a certain $a \varepsilon A$, the number $b^*=c-a$, as we know, occurs in B^*. We have $c=a+b^* \varepsilon A+B^* \subset A+B_1$. Therefore $C_1 \subset A+B_1$. We also proved above that $A+B_1 \subset C_1$. Consequently $C_1=A+B_1$.

Now recall that according to our assumption, $n \notin C$. It is easy to see—and this is important—that the number n does not appear in the extension C_1. For if we had $n \varepsilon C^*$, we could, by the definition of C^*, put $c'=n$ in equation (9), which would yield $c=a+\beta_0 \varepsilon A+B=C$, whereas $c \notin C$ according to (9).

If the extended sequence C_1 is not yet normal, then, because of $A+B_1=C_1$ and $n \notin C_1$, the sets A, B_1, and C_1 form a triple with all the properties of the triple A, B, C that are necessary for a new canonical extension. We take a new basis β_1 of this extension, define the complementary sets B_1^*, C_1^* as before, put

$$B_1 \cup B_1^*=B_2, \quad C_1 \cup C_1^*=C_2,$$

and are able to assert once more that $A+B_2=C_2$ and $n \notin C_2$. It is evident that this process can be continued until one of the extensions C_h proves to be normal. Obviously this case must certainly take place, because in every extension we add new numbers to the sets B_μ and C_μ without overstepping the bounds of the segment $(0, n)$.

In this way we obtain the finite sequences of sets

$$B=B_0 \subset B_1 \subset \ldots \subset B_h,$$
$$C=C_0 \subset C_1 \subset \ldots \subset C_h,$$

where every $B_{\mu+1}$ (respectively $C_{\mu+1}$) contains numbers which do not appear in B_μ (C_μ) and which go to make up the set B_μ^* (C_μ^*), so that

$$B_{\mu+1} = B_\mu \cup B_\mu^*, \quad C_{\mu+1} = C_\mu \cup C_\mu^* \quad (0 \leqq \mu \leqq h-1).$$

We denote by β_μ the basis of the extension which carries (B_μ, C_μ) into $(B_{\mu+1}, C_{\mu+1})$. We have

$$A + B_\mu = C_\mu, \quad n \notin C_\mu \quad (0 \leqq \mu \leqq h).$$

Finally, the set C_h is normal, whereas the sets C_μ $(0 \leqq \mu \leqq h-1)$ are not.

§7
PROPERTIES OF THE CANONICAL EXTENSIONS

We shall now formulate and prove in the form of three lemmas those properties of the canonical extensions which are needed later. Only Lemma 3 will have further application; Lemmas 1 and 2 are required solely for the proof of Lemma 3.

Lemma 1. $\beta_\mu > \beta_{\mu-1}$ $(1 \leqq \mu \leqq h-1)$; *i. e., the bases of successive canonical extensions form a monotonically increasing sequence.*

In fact, since $\beta_\mu \in B_\mu = B_{\mu-1} \cup B_{\mu-1}^*$, either $\beta_\mu \in B_{\mu-1}^*$ or $\beta_\mu \in B_{\mu-1}$. If $\beta_\mu \in B_{\mu-1}^*$, then β_μ is of the form

$$\beta_\mu = \beta_{\mu-1} + n - c,$$

where $c \in C_{\mu-1}^* \subset C_\mu$ and therefore $c < n$, so that $\beta_\mu > \beta_{\mu-1}$, and Lemma 1 is proved. If $\beta_\mu \in B_{\mu-1}$, however, then by the definition of the number β_μ there exist integers $a \in A$, $c \notin C_\mu$, $c' \notin C_\mu$ such that

$$c + c' - n = a + \beta_\mu \in C_\mu.$$

But for $\beta_\mu \in B_{\mu-1}$, we have

$$c + c' - n = a + \beta_\mu \in A + B_{\mu-1} = C_{\mu-1}, \tag{10}$$

where $c \notin C_{\mu-1}$, $c' \notin C_{\mu-1}$. Hence, because of the minimal property of $\beta_{\mu-1}$, $\beta_\mu \geqq \beta_{\mu-1}$. If $\beta_\mu = \beta_{\mu-1}$, it would follow from (10) and the definition of the set $C_{\mu-1}^*$ that

$$c \in C_{\mu-1}^* \subset C_\mu, \quad c' \in C_{\mu-1}^* \subset C_\mu.$$

Both are false, however, and therefore $\beta_\mu > \beta_{\mu-1}$.

In the sequel we shall denote by m the smallest positive integer which does not appear in C_h.

LEMMA 2. *If* $c \in C^*_\mu$ $(0 \leq \mu \leq h-1)$ *and* $n-m<c<n$, *then* $c>n-m+\beta_\mu$. *That is, all numbers c of the set* C^*_μ *which lie in the interval* $n-m<c<n$ *are embedded in that part of this segment which is characterized by the inequalities* $n-m+\beta_\mu<c<n$.

We have to show that

$$c+m-n>\beta_\mu.$$

It follows from $n-m<c<n$ that

$$0<m+c-n<m.$$

Therefore, by the definition of the number m,

$$m+c-n \in C_h.$$

Now

$$C_h = C_\mu \cup C^*_\mu \cup C^*_{\mu+1} \cup \ldots \cup C^*_{h-1}.$$

We consider two cases.

1) If $m+c-n \in C_\mu$, then

$$m+c-n=a+b_\mu, \qquad a \in A,\ b_\mu \in B_\mu.$$

But $m \notin C_\mu$ and $c \notin C_\mu$ (the latter because $c \in C^*_\mu$). Therefore because of the minimal property of β_μ we must have $b_\mu \geqq \beta_\mu$. If $b_\mu=\beta_\mu$ it would follow from the definition of the set C^*_μ that $m \in C^*_\mu$, which is false because $C^*_\mu \subset C_{\mu+1} \subset C_h$ and $m \notin C_h$. Consequently $b_\mu>\beta_\mu$, so that

$$m+c-n=a+b_\mu \geqq b_\mu>\beta_\mu,$$

and Lemma 2 is proved.

2) If $c'=m+c-n \in C^*_\nu$ $(\mu \leqq \nu \leqq h-1)$, then, by the definition of the set C^*_ν, c' satisfies an equation of the form (9),

$$c'-a=\beta_\nu+n-c'',$$

where $a \in A$, $c'' \in C^*_\nu$. Hence $c' \geqq c'-a>\beta_\nu \geqq \beta_\mu$ (where the last inequality is given by Lemma 1), and Lemma 2 is again proved.

LEMMA 3. *We have*

$$C^*_\mu(n)-C^*_\mu(n-m)=B^*_\mu(m-1) \qquad (0\leq\mu\leq h-1).$$

That is, the number of integers $c\in C^*_\mu$ *in the segment* $n-m<c<n$ *is exactly the same as the number of integers* $b\in B^*_\mu$ *in the segment* $0<b<m$ *(of the same length).*

Let us examine the relation

$$b=\beta_\mu+n-c. \tag{11}$$

By the very definition of the sets B^*_μ and C^*_μ, $c\in C^*_\mu$ implies $b\in B^*_\mu$, and conversely. If, in addition, $n-m+\beta_\mu<c<n$, then $\beta_\mu<b<m$, and conversely. Hence

$$C^*_\mu(n)-C^*_\mu(n-m+\beta_\mu)=B^*_\mu(m-1)-B^*_\mu(\beta_\mu).$$

By Lemma 2, $C^*_\mu(n-m+\beta_\mu)=C^*_\mu(n-m)$. On the other hand, every $b\in B^*_\mu$ can be expressed in the form (11), where $c<n$; b therefore exceeds β_μ, and consequently $B^*_\mu(\beta_\mu)=0$. It follows that

$$C^*_\mu(n)-C^*_\mu(n-m)=B^*_\mu(m-1), \qquad \text{Q.E.D.}$$

§ 8

PROOF OF THE FUNDAMENTAL LEMMA

It is very easy now to prove the fundamental lemma by proceeding from the results in §5 and appealing to Lemma 3 which was just proved.

If we apply the result of §5 in the form of inequality (7) to the sequences A, B_h, and C_h (which is permissible because of the normality of C_h), we find that

$$C_h(n)-C_h(n-m)\geq A(m)+B_h(m), \tag{12}$$

where m is the smallest positive integer which does not occur in C_h. Obviously $m\notin A$ and $m\notin B_h$, so that we may write $A(m-1)$ and $B_h(m-1)$ instead of $A(m)$ and $B_h(m)$, respectively.

We have

$$C_h=C\cup C^*\cup C^*_1\cup\ldots\cup C^*_{h-1},$$
$$B_h=B\cup B^*\cup B^*_1\cup\ldots\cup B^*_{h-1},$$

where the sets appearing in any one of these two unions are mutually exclusive, so that

$$C_h(n)-C_h(n-m)=C(n)-C(n-m)+\sum_{\mu=0}^{h-1}\{C_\mu^*(n)-C_\mu^*(n-m)\},$$

$$B_h(m)=B_h(m-1)=B(m-1)+\sum_{\mu=0}^{h-1}B_\mu^*(m-1);$$

we have of course put $C_0^*=C^*$, $B_0^*=B^*$. On account of (12) it follows that

$$C(n)-C(n-m)+\sum_{\mu=0}^{h-1}\{C_\mu^*(n)-C_\mu^*(n-m)\}$$

$$\geqq A(m)+B(m-1)+\sum_{\mu=0}^{h-1}B_\mu^*(m-1).$$

By Lemma 3, however,

$$C_\mu^*(n)-C_\mu^*(n-m)=B_\mu^*(m-1) \qquad (0\leqq\mu\leqq h-1),$$

so that the preceding inequality becomes

$$C(n)-C(n-m)\geqq A(m)+B(m-1)=A(m)+B(m)\geqq(\alpha+\beta)m,$$

which proves the fundamental lemma.

As we saw in § 4, this also completes the proof of Mann's theorem which solves the fundamental metric problem of additive number theory.

Doesn't Artin and Scherk's construction have the stamp of a magnificent masterpiece? I find the outstanding combination of structural finesse and the extremely elementary form of the method especially attractive.

CHAPTER III

AN ELEMENTARY SOLUTION OF WARING'S PROBLEM

§1

You will recall the theorem of Lagrange, which was discussed at the beginning of the preceding chapter. It says that every natural number can be expressed as the sum of at most four squares. I also showed you that this theorem could be stated in entirely different terms: If four sequences, each identical with

$$(A_2) \qquad 0, 1^2, 2^2, \ldots, k^2, \ldots,$$

are added together, the resulting sequence contains all the natural numbers. Or even more briefly, the sequence (A_2) is a basis (of the sequence of natural numbers) of order four. I also mentioned that, as had been shown later, the sequence of cubes

$$(A_3) \qquad 0, 1^3, 2^3, \ldots, k^3, \ldots$$

was a basis of order nine. All these facts lead in a natural manner to the hypothesis that, for an arbitrary natural number n, the sequence

$$(A_n) \qquad 0, 1^n, 2^n, \ldots, k^n, \ldots$$

is a basis (whose order of course depends on n). This conjecture was also actually propounded by Waring as early as the eighteenth century. The problem proved to be very difficult, however, and it was not until the beginning of the present century that the universal validity of Waring's hypothesis was demonstrated, by Hilbert (1909). Hilbert's proof is not only ponderous in its formal aspect and based on complicated analytical theories (multiple integrals), but also lacks transparency in conceptual respects. The eminent French mathematician Poincaré wrote in his survey of Hilbert's creative scientific work, that once the basic motivations behind this proof were understood, arithmetical results of great importance would probably flow forth as from a cornucopia. In a certain sense he was right. Ten to fifteen years later, new proofs of Hilbert's theorem were furnished by Hardy and Littlewood in England and by I. M. Vinogradov in the USSR. These proofs were again analytic and formally unwieldy, but differed favorably from Hilbert's proof in their clarity of method and their conceptual simplicity, which left nothing to be desired. In fact, because of this, both methods became mighty scources of new arithmetical theorems.

But when our science is concerned with such a completely elementary problem as Waring's problem, it invariably attempts to find a solution which requires no concepts or methods transcending the the limits of elementary arithmetic. The search for such an elementary proof of Waring's hypothesis is the third problem which I should like to tell you about. Such a fully elementary proof of Hilbert's theorem was first obtained in 1942, by the young Soviet scholar Y. V. Linnik.

You are already accustomed to the fact that "elementary" does not mean "simple". The elementary solution of Waring's problem discovered by Linnik is, as you will see, not very simple either, and it will take considerable effort on your part to understand and digest it. I shall endeavor to make this task as easy as possible for you through my exposition. But you must remember that in mathematics (as probably in any other science) the assimilation of anything really valuable and significant involves trying labor.

The ideas of Schnirelmann which I described to you in the beginning of the second chapter play an essential role in Linnik's proof. You will recall (I mentioned it at that time) how Schnirelmann proved his famous theorem that the sequence P consisting of zero, unity, and all the primes, is a basis of the sequence of natural numbers: He showed that the sequence $P+P$ has a positive density. This immediately yields the assertion, however, because, according to the general theorem of Schnirelmann which we proved on pp.24-25, every sequence of positive density is a basis of the sequence of natural numbers. The same method also lies at the basis of the proof of Hilbert's theorem discovered by Linnik. It all boils down to the proof that the sum of a sufficiently large number of sequences (A_n) is a sequence of positive density. As soon as this is accomplished, we can, by virtue of the same general theorem of Schnirelmann, regard Hilbert's theorem as proved.

§2
THE FUNDAMENTAL LEMMA

If we add together k sequences, identical with A_n, according to the rule in Chapter II, we evidently obtain a sequence $A_n^{(k)}$ which contains zero and all those natural numbers which can be expressed as a sum of at most k summands of the form x^m, where x is an arbitrary natural number. In other words, the number m belongs to the sequence $A_n^{(k)}$, if the equation

$$x_1^n + x_2^n + \dots + x_k^n = m \tag{1}$$

is solvable in nonnegative integers x_i $(1 \leq i \leq k)$. As we saw in §1, the problem is to show that, for sufficiently large k, the sequence $A_n^{(k)}$ has a positive density.

For preassigned k and m, equation (1) in general can be solved in several different ways. In the sequel we shall denote by $r_k(m)$ the number of these ways, i.e., the number of systems of nonnegative integers $x_1, x_2, \dots, x_k$ which satisfy equation (1). It is clear that the number m occurs in $A_n^{(k)}$ if and only if $r_k(m) > 0$.

In the following, we shall assume the number n to be given and fixed, and shall therefore call all numbers which depend only upon n, constants. Such constants will be denoted by the letter c or $c(n)$, where such a constant c may have different values in different parts of our discussion, provided merely that these values are constants. Perhaps you are rather unused to such "freedom" of notation, but you will soon become familiar with it. It has proved to be very convenient, and appears more and more frequently in modern research.

FUNDAMENTAL LEMMA. *There exist a natural number $k = k(n)$, depending only on n, and a constant c, such that, for an arbitrary natural number N,*

$$r_k(m) < c N^{(k/n)-1} \quad (1 \leq m \leq N). \tag{2}$$

Once more, as in the preceding chapter, we are faced with two problems: first, to prove the fundamental lemma, and second, to draw from the fundamental lemma the conclusion that we need, viz., that the sequence $A_n^{(k)}$ has a positive density. This time again the second problem is considerably easier than the first, and we shall therefore begin with the second problem.

It follows immediately from the definition of the number $r_k(m)$, that the sum

$$r_k(0)+r_k(1)+\dots+r_k(N)=R_k(N)$$

represents the number of systems $(x_1, x_2, \dots, x_k)$ of k nonnegative integers for which

(3) $$x_1^n+x_2^n+\dots+x_k^n \leqq N.$$

Every group of numbers for which

$$0 \leqq x_i \leqq (N/k)^{1/n} \qquad (1 \leqq i \leqq k),$$

obviously satisfies this condition. To satisfy these inequalities, every x_i can evidently be chosen in more than $(N/k)^{1/n}$ different ways $(x_i=0, 1, \dots, [(N/k)^{1/n}])$.* After an arbitrary choice of this sort, the numbers $x_1, x_2, \dots, x_k$ may be combined, and so we have more than $(N/k)^{k/n}$ different possibilities for choosing the complete system of integers x_i $(1 \leqq i \leqq k)$ so as to satisfy condition (3). This shows that

(4) $$R_k(N) \geqq (N/k)^{k/n}.$$

We assume that the fundamental lemma has been shown to be correct, and that inequality (2) is satisfied for an arbitrary N. We now have to verify that inequality (2) is consistent with inequality (4) which we proved, only if the sequence $A_n^{(k)}$ has a positive density. The idea behind the following deduction is very simple: In the sum $R_k(N)$, only those summands $r_k(m)$ are different from zero, for which m occurs in $A_n^{(k)}$. If $A_n^{(k)}$ had density zero, then for large N the number of such summands would be relatively small; because of (2), however, every summand cannot be very large. Their sum $R_k(N)$, therefore, would also be relatively small, whereas according to (4) it must be rather large.

* $[a]$ denotes the largest integer $\leqq a$.

It remains to carry out the calculations. Suppose that $d(A_n^{(k)})=0$. Then, for an arbitrary small $\epsilon>0$ and a suitably chosen N,

$$A_n^{(k)}(N)<\epsilon N.$$

Here the number N may be assumed to be arbitrarily large, because $A_n^{(k)}$ (for an arbitrary k) contains the integer 1 (bear in mind Problem 6 on p. 22, which you solved). Applying the estimate (2) we get

$$R_k(N)=\sum_{m=0}^{n} r_k(m)=r_k(0)+\sum_{m=1}^{N} r_k(m)<1+cN^{(k/n)-1}A_n^{(k)}(N)<1+c\epsilon N^{(k/n)},$$

and hence, for sufficiently large N,

$$R_k(N)<2c\epsilon N^{k/n}.$$

For sufficiently small ϵ,

$$2c\epsilon<(1/k)^{k/n},$$

so that

$$R_k(N)<(N/k)^{k/n},$$

which contradicts (4). Therefore we must have

$$d(A_n^{(k)})>0.$$

But, as we already know, this proves Hilbert's theorem.

You see how simply it all comes out. But we still have to prove the fundamental lemma, and to do this we shall have to travel a long and difficult road, as in the preceding chapter.

§3

LEMMAS CONCERNING LINEAR EQUATIONS

We shall have to go far back. It will therefore be well for you to forget completely for a while the problem which has been posed. I shall call your attention to it when we return to it later.

Right now, however, we have to find some estimates for the number of solutions of systems of linear equations. The lemmas of this paragraph, moreover, are perhaps also of intrinsic interest, independent of the problem for whose solution they are required here.

LEMMA 1. *In the equation*

(5) $$a_1z_1 + a_2z_2 = m,$$

let a_1, a_2, m *be integers with* $|a_2| \leqq |a_1| \leqq A$, *and let* a_1 *and* a_2 *be relatively prime. Then the number of solutions of equation* (5) *satisfying the inequalities* $|z_1| \leqq A$, $|z_2| \leqq A$, *does not exceed* $3A/|a_1|$.

Proof: We may assume that $a_1 > 0$, because otherwise we have merely to replace z_1 by $-z_1$ in every solution.

Let $\{z_1, z_2\}$ and $\{z_1', z_2'\}$ be two different solutions of equation (5). Then from

$$a_1z_1 + a_2z_2 = m,$$
$$a_1z_1' + a_2z_2' = m$$

we get

$$a_2(z_2' - z_2) = a_1(z_1 - z_1')$$

by subtraction. Accordingly the left-hand side of this equation must be divisible by a_1. But* $(a_1, a_2) = 1$, and consequently $z_2' - z_2$ must be divisible by a_1. Now $z_2' \neq z_2$, and therefore $|z_2' - z_2|$, as a multiple of a_1, is not smaller than a_1. Thus, for two distinct solutions $\{z_1, z_2\}$ and $\{z_1', z_2'\}$ of equation (5), we must have $|z_2' - z_2| \geqq a_1$.

In every solution $\{z_1, z_2\}$ of equation (5), let us agree to call z_1 the first member and z_2 the second. It is obvious that the number of solutions of equation (5) which satisfy the conditions $|z_1| \leqq A$, $|z_2| \leqq A$, is not more than the number t of second members which occur in the interval $\langle -A, A \rangle$. Since we have proved that two such second members are at least the distance a_1 apart, the difference between the largest and smallest second members occurring in the interval $\langle -A, A \rangle$ is at least $a_1(t-1)$. On the other hand, this difference does not exceed $2A$, so that

$$a_1(t-1) \leqq 2A,$$
$$(t-1) \leqq 2A/a_1,$$
$$t \leqq (2A/a_1) + 1 \leqq 3A/a_1$$

(because, by assumption, $a_1 \leqq A$, and therefore $1 \leqq A/a_1$). This proves Lemma 1.

LEMMA 2. *In the equation*

(6) $$a_1z_1 + a_2z_2 + \ldots + a_lz_l = m,$$

* (a_1, a_2) denotes the greatest common divisor of the integers a_1 and a_2.

*let the a_i and m be integers satisfying the conditions**

$$|a_i| \leqq A \quad (1 \leqq i \leqq l), \qquad (a_1, a_2, \ldots, a_l) = 1.$$

Then the number of solutions of equation (6) *satisfying the inequalities* $|z_i| \leqq A$ $(1 \leqq i \leqq l)$, *does not exceed*

$$c(l)\, A^{l-1}/H,$$

where H is the largest of the numbers $|a_1|, |a_2|, \ldots, |a_l|$, *and* $c(l)$ *is a constant depending only on l.*

Proof: If $l = 2$, Lemma 2 obviously becomes Lemma 1 (with $c(2) = 3$). Accordingly Lemma 2 is already verified for $l = 2$. We shall therefore assume that $l \geq 3$ and that the truth of Lemma 2 has already been established for the case of $l-1$ unknowns. Since the numbering is unimportant, we may assume that $|a_l|$ is the largest of the numbers $|a_1|, |a_2|, \ldots, |a_l|$, i. e., $H = |a_l|$.

There are two cases to consider.

1) $a_1 = a_2 = \ldots = a_{l-1} = 0$. Since $(a_1, a_2, \ldots, a_l) = 1$, we have $|a_l| = H = 1$, so that the given equation is of the form $\pm z_l = m$. In this equation each of the unknowns $z_1, z_2, \ldots, z_{l-1}$ can obviously assume an arbitrary integral value in the interval $\langle -A, A \rangle$, and hence at most $2A + 1 \leq 3A$ values all told. As for z_l, however, it can assume at most one value. Consequently the number of solutions of the given equation satisfying the inequalities $|z_i| \leqq A$ $(1 \leqq i \leqq l)$, does not exceed

$$(3A)^{l-1} = c(l)\, A^{l-1} = c(l)\, A^{l-1}/H,$$

which proves Lemma 2 for this case.

2) If at least one of the numbers $a_1, a_2, \ldots, a_{l-1}$ is different from zero, then

$$(a_1, a_2, \ldots, a_{l-1}) = \delta$$

exists. Let us denote by H' the largest of the numbers

$$|a_i|/\delta \qquad (1 \leqq i \leqq l-1).$$

Suppose now that the numbers $z_1, z_2, \ldots, z_l$ satisfy the given equation (6) and the inequalities $|z_i| \leqq A$ $(1 \leqq i \leqq l)$. We set

(7) $$(a_1/\delta)\, z_1 + (a_2/\delta) z_2 + \ldots + (a_{l-1}/\delta)\, z_{l-1} = m',$$

*$(a_1, a_2, \ldots, a_l)$ denotes the greatest common divisor of the integers in parentheses.

and hence

$$a_1z_1+a_2z_2+\dots+a_{l-1}z_{l-1}=\delta m'.$$

Then obviously

$$\delta m'+a_lz_l=m \tag{8}$$

and

$$|\delta m'|\leqq \sum_{i=1}^{l-1} |a_i|\,|z_i|\leqq l\,\delta\, H'A,$$

which implies that

$$|m'|\leqq lH'A.$$

Thus, if the numbers $z_1, z_2, \dots, z_l$ satisfy equation (6) and the inequalities $|z_i|\leqq A$ $(1\leqq i\leqq l)$, then the integer m' exists, which, with these numbers, satisfies equations (7) and (8), where $|m'|\leqq lH'A$. But in equation (8) clearly $\delta\leqq|a_l|$ and $(\delta, a_l)=1$ (otherwise we should have $(a_1, a_2, \dots, a_{l-1}, a_l)>1$). Hence, by Lemma 1, the number of solutions of equation (8) (in the unknowns m', z_l), for which $|m'|\leqq lH'A$, $|z_l|\leqq A<lH'A$, does not exceed $3lH'A/|a_l|$. For the same m', equation (7), according to Lemma 2 for equations in $l-1$ unknowns, has at most $c(l)A^{l-2}/H'$ solutions in integral z_i with $|z_i|\leqq A$.

It is evident, from what has been said, that the number of solutions $\{z_1, z_2, \dots, z_l\}$ of equation (6) which satisfy the inequalities $|z_i|\leqq A$ $(1\leqq i\leqq l)$, does not exceed

$$\begin{aligned}(3lH'A/|a_l|)\,c(l)A^{l-2}/H' &= c(l)A^{l-1}/|a_l|\\ &= c(l)A^{l-1}/H,\end{aligned}$$

which completes the proof of Lemma 2.*

We shall now investigate the totality of equations of the form

$$a_1z_1+a_2z_2+\dots+a_lz_l=0, \tag{9}$$

where $|a_i|\leqq A$ $(1\leqq i\leqq l)$ and, as always, all a_i are integers. Let B be a positive number whose relation to the number A is described by the inequalities $1\leqq A\leqq B\leqq c(l)A^{l-1}$, and let $l>2$. We now want to estimate

*You have probably noticed that in the last chain of equations the symbol $c(l)$ occurred in different places with different meanings. On p. 39 I prepared you for such a use of this symbol.

the sum of the numbers of solutions z_i, $|z_i| \leqq B$ $(1 \leqq i \leqq l)$ of all the equations (9) of this family.

1° First let us make a separate examination of equation (9) for $a_1 = a_2 = \ldots = a_l = 0$ (it is a member of our family) and estimate the number of its solutions which satisfy the inequalities $|z_i| \leqq B$ $(1 \leqq i \leqq l)$. Our equation is obviously satisfied by an arbitrary system of numbers $z_1, z_2, \ldots, z_l$, and we have merely to calculate how many such systems exist which satisfy the inequalities $|z_1| \leqq B$, $|z_2| \leqq B$, $\ldots, |z_l| \leqq B$. Since the interval $\langle -B, +B \rangle$ contains at most $2B+1$ integers, each z_i can assume at most $2B+1$ different values. Consequently the number of systems $\{z_1, z_2, \ldots, z_l\}$ of the type in which we are interested does not exceed $(2B+1)^l \leqq (3B)^l = c(l)B^l$. By our hypothesis, however, $B \leqq c(l)A^{l-1}$, so that $c(l)B^l = c(l)B^{l-1}B \leqq c(l)(AB)^{l-1}$. Hence, for the case where $a_1 = a_2 = \ldots = a_l = 0$, equation (9) has at most $c(l)(AB)^{l-1}$ solutions of the type we are interested in.

2° Even if only one of the coefficients a_i is different from zero, the greatest common divisor of these coefficients, $(a_1, a_2, \ldots, a_l) = \delta$, exists. Suppose first that $\delta = 1$, and let H be the largest of the numbers $|a_i|$ $(i = 1, 2, \ldots, l)$. Clearly H is one of the integers in the interval $\langle 1, A \rangle$. Hence, H is either between A and $A/2$, or between $A/2$ and $A/4$, or between $A/4$ and $A/8$, etc. It is therefore possible to find an integer $m \geqq 0$ such that

$$A/2^{m+1} < H \leqq A/2^m. \tag{10}$$

According to Lemma 2, for an equation of the form (9) in which $\delta = 1$ and H satisfies the inequalities (10), the number of solutions z_i, $|z_i| \leqq B$, does not exceed

$$c(l)B^{l-1}/H \leqq c(l)B^{l-1}/(A/2^{m+1}) = c(l)B^{l-1}2^m/A.$$

On the other hand, it follows from (10) that

$$|a_i| \leqq A/2^m \quad (1 \leqq i \leqq l). \tag{11}$$

Consequently the number of equations of type (9) for which the inequalities (10) are satisfied is at most equal to the number of equations of the same type which satisfy the conditions (11), i. e., at most

$$(2(A/2^m)+1)^l \leqq (3A/2^m)^l = c(l)A^l 2^{-ml}.$$

Thus the sum of the numbers of solutions $|z_i| \leqq B$ of all such equations of type (9) for which $\delta = 1$ and $A2^{-(m+1)} < H \leqq A2^{-m}$, doesn't exceed

$$(c(l)B^{l-1}2^m/A) \cdot c(l)A^l 2^{-ml} = c(l)(AB)^{l-1}2^{-(l-1)m}.$$

Summing this estimate over all $m \geqq 0$, we reach the following conclusion: The sum of the numbers of solutions $|z_i| \leqq B$ of all equations (9) for which $|a_i| \leqq A$ $(1 \leqq i \leqq l)$ and $\delta = 1$ is at most

$$c(l)(AB)^{l-1}.$$

3° It remains for us to figure out the numbers of solutions of the required type for equations with $\delta > 1$. In this case equation (9) is evidently synonymous with the equation

$$(a_1/\delta)z_1 + (a_2/\delta)z_2 + \ldots + (a_l/\delta)z_l = 0,$$

where only

$$(a_1/\delta, a_2/\delta, \ldots, a_l/\delta) = 1$$

and the number A has to be replaced by the number A/δ. As we saw in 2°, the sum of the numbers of solutions $|z_i| \leqq B$ of all such equations, for a given, fixed δ, does not exceed*

$$c(l)(A\delta^{-1} \cdot B)^{l-1} = c(l)(AB)^{l-1}\delta^{-(l-1)}.$$

Clearly now we have merely to sum this expression over all the possible values of δ $(1 \leqq \delta \leqq A)$.

Thus we find that the sum of the numbers of required solutions of all equations of the form (9), where $|a_i| \leqq A$ $(1 \leqq i \leqq l)$ and not all a_i are equal to zero, does not exceed the value

$$c(l)(AB)^{l-1} \sum_{\delta=1}^{A} \delta^{-(l-1)} < c(l)(AB)^{l-1} \cdot \frac{l-1}{l-2} = c(l)(AB)^{l-1}.$$

[To obtain the first relation we employ the inequality

$$\sum_{n=1}^{A} (1/n^{q+1}) < (q+1)/q,$$

*Since instead of A we now have to take the smaller number A/δ, it is conceivable that the assumed condition $B \leqq c(l)A^{l-1}$ is violated. You can verify, however, without any trouble, that we made no use of this assumption in Case 2°, and that the result in 2° therefore does not depend on it.

which is valid for an arbitrary natural number q and for an arbitrary $A \geqq 1$ (we denote by q the number $l-2$, which is positive because we assumed that $l>2$). Here is a simple proof: For $n \geqq 1$ we have

$$\begin{aligned} n^{-q}-(n+1)^{-q} &= \{(n+1)^q-n^q\}/n^q(n+1)^q \\ &= (n^q+qn^{q-1}+\ldots+1-n^q)/n^q(n+1)^q \\ &\geqq qn^{q-1}/n^q(n+1)^q > q/(n+1)^{q+1}, \end{aligned}$$

and hence

$$(n+1)^{-(q+1)} < q^{-1}\{n^{-q}-(n+1)^{-q}\}.$$

By substituting successively $n=1, 2, \ldots, A-1$ in this inequality and adding all the resulting inequalities together we find that

$$\sum_{n=2}^{A} n^{-(q+1)} < q^{-1}(1-A^{-q}) < 1/q,$$

which implies that

$$\sum_{n=1}^{A} n^{-(q+1)} < 1+(1/q) = (q+1)/q, \qquad \text{Q.E.D.]}$$

Comparing this with the result in 1°, where we obtained an estimate for the case $a_1=a_2=\ldots=a_l=0$, we reach the following conclusion:

LEMMA 3. *Let $l>2$ and $1 \leqq A \leqq B \leqq c(l)A^{l-1}$. Then the sum of the numbers of solutions $|z_i| \leqq B$ $(1 \leqq i \leqq l)$ of all equations of the form*

$$a_1z_1+a_2z_2+\ldots+a_lz_l=0, \tag{9}$$

where $|a_i| \leqq A$ $(1 \leqq i \leqq l)$, does not exceed

$$c(l)(AB)^{l-1}.$$

§4
TWO MORE LEMMAS

Before proceeding to prove the fundamental lemma, we have to derive two more lemmas of a special type. They are both very simple, in idea as well as in form, and yet their assimilation might cause you some difficulty because they are concerned with the enumeration of all possible combinations, whose construction is rather involved. The difficulty with such an abstract combinatorial problem is that it is hard to put it in mathematical symbols: one has to express more in words than in signs. This is of course a difficulty of presentation, however, and not of the subject itself, and I shall take pains to outline all questions that arise, and their solution, as concretely as possible.

We shall denote by A a finite complex (i. e., collection) of numbers, not all of which are necessarily distinct. If the number a occurs λ times in the complex A, we shall say that its multiplicity is λ. Let $a_1, a_2, \ldots, a_r$ be the distinct numbers which appear in A, and let $\lambda_1, \lambda_2, \ldots, \lambda_r$ be their respective multiplicities (because the complex A contains all together $\sum_{i=1}^{r}\lambda_i$ numbers). Let B be another complex of the same type, which consists of the distinct numbers b_1, $b_2, \ldots, b_s$ with the respective multiplicities $\mu_1, \mu_2, \ldots, \mu_s$.

Let us investigate the equation

$$x+y=c, \tag{12}$$

where c is a given number and x and y are unknowns. We are interested in such solutions $\{x, y\}$ of this equation in which x is one of the numbers of the complex A (abbreviated $x \varepsilon A$) and y is one of the numbers of the complex B ($y \varepsilon B$). If the numbers $x=a_i$ and $y=b_k$ satisfy equation (12), this yields $\lambda_i\mu_k$ solutions of the required kind, because any one of the λ_i "specimens" of the number a_i, which occur in the complex A, can be combined with an arbitrary one of the μ_k specimens of the number b_k appearing in the complex B. But we have* $\lambda_i\mu_k \leqq \frac{1}{2}(\lambda_i^2+\mu_k^2)$. Therefore the number of such solutions of quation (12), where $x=a_i$, $y=b_k$, is not greater than $\frac{1}{2}(\lambda_i^2+\mu_k^2)$. It follows that the number of all solutions $x \varepsilon A$, $y \varepsilon B$ of equation (12) is not more than the sum $\sum\frac{1}{2}(\lambda_i^2+\mu_k^2)$. Here the summation is over all pairs of indices $\{i, k\}$ for which $a_i+b_k=c$. Our sum is enlarged if we sum λ_i^2 over all i and μ_k^2 over all k (because every b_k can be combined with at most one a_i.) It finally follows, therefore, that the number of solutions $x \varepsilon A$, $y \varepsilon B$ of equation (12) does not exceed the number

$$\frac{1}{2}\Big(\sum_{i=1}^{r}\lambda_i^2+\sum_{k=1}^{s}\mu_k^2\Big).$$

*"The geometric mean is not greater than the arithmetic mean". Here is the simplest proof:

$$0 \leqq (\lambda_i-\mu_k)^2=\lambda_i^2+\mu_k^2-2\lambda_i\mu_k, \text{ and hence } 2\lambda_i\mu_k \leqq \lambda_i^2+\mu_k^2.$$

On the other hand, let us consider the equation

(13) $$x-y=0$$

and calculate the number of its solutions $x \varepsilon A$, $y \varepsilon A$. Clearly every such solution is of the form $x=y=a_i$ $(1 \leqq i \leqq r)$. For a given i we obtain λ_i^2 solutions, because the numbers x and y can coincide, independently of one another, with any one of the λ_i specimens of the number a_i appearing in A. Accordingly the total number of solutions $x \varepsilon A$, $y \varepsilon A$ of equation (13) is equal to $\sum_{i=1}^{r} \lambda_i^2$. In exactly the same way we find, of course, that the number of solutions $x \varepsilon B$, $y \varepsilon B$ of the same equation is equal to $\sum_{k=1}^{s} \mu_k^2$. If we compare these results with the one found above, we reach the following conclusion:

LEMMA 4. *The number of solutions of the equation*

$$x+y=c, \quad x \varepsilon A,\ y \varepsilon B$$

does not exceed half the sum of the numbers of solutions of the equations

$$x-y=0, \quad x \varepsilon A,\ y \varepsilon A$$

and

$$x-y=0, \quad x \varepsilon B,\ y \varepsilon B.$$

For the special case in which the complexes A and B coincide we obtain the following

COROLLARY. *The number of solutions of the equation*

$$x+y=c, \quad x \varepsilon A,\ y \varepsilon A$$

does not exceed the number of solutions of the equation

$$x-y=0, \quad x \varepsilon A,\ y \varepsilon A.$$

Now let k and s be two arbitrary natural numbers. We put $k \cdot 2^s = l$, and investigate the equation

$$x_1+x_2+\ldots+x_l=c.$$

Let $A_1, A_2, \ldots, A_l$ be finite complexes of numbers. Suppose that the complex A_i $(1 \leqq i \leqq l)$ consists of the distinct numbers $a_{i1}, a_{i2}, \ldots$

with the respective multiplicities $\lambda_{i1}, \lambda_{i2}, \ldots$. We are interested in the number of solutions of the equation

$$(14) \qquad x_1+x_2+\ldots+x_l=c, \quad x_i \varepsilon A_i \ (1 \leqq i \leqq l).$$

If we set

$$x_1+x_2+\ldots+x_{l/2}=x, \quad x_{(l/2)+1}+\ldots+x_l=y$$

($l/2$ is of course an integer), then the given equation can be written in the form

$$x+y=c,$$

and Lemma 4, which we have just proved, can be applied to it. We have only to find out to which complexes the numbers x and y belong. Since $x_i \varepsilon A_i$ $(1 \leqq i \leqq l)$, x can be an arbitrary number of the form $z_1+z_2+\ldots+z_{l/2}$, where $z_i \varepsilon A_i$ $(1 \leqq i \leqq l/2)$. Similarly y can be an arbitrary number of the same form, where, however, $z_i \varepsilon A_{(l/2)+i}$ $(1 \leqq i \leqq l/2)$.

Hence, by Lemma 4, the number of solutions of equation (14) does not exceed half the sum of the numbers of solutions of the equation

$$(15) \qquad x-y=0$$

under the following two hypotheses:

1)
$$x=z_1+z_2+\ldots+z_{l/2},$$
$$y=z_1'+z_2'+\ldots+z_{l/2}',$$

where

$$(16) \qquad z_i \varepsilon A_i, \quad z_i' \varepsilon A_i \quad (1 \leqq i \leqq l/2);$$

2) x and y have the same form, but

$$(17) \qquad z_i \varepsilon A_{(l/2)+i}, \quad z_i' \varepsilon A_{(l/2)+i} \ (1 \leqq i \leqq l/2).$$

In both cases equation (15) may be rewritten in the form

$$(18) \qquad (z_1-z_1')+(z_2-z_2')+\ldots+(z_{l/2}-z_{l/2}')=0.$$

We conclude therefore that the number of solutions of equation (14) does not exceed half the sum of the numbers of solutions of equation

(18) under the hypotheses (16) and (17), i. e., it does not exceed half the sum of the numbers of solutions of the equations

(18a) $\sum_{i=1}^{l/2}(z_i - z_i') = 0, \quad z_i \varepsilon A_i, \qquad z_i' \varepsilon A_i \qquad (1 \leqq i \leqq l/2)$

and

(18 b) $\sum_{i=1}^{l/2}(z_i - z_i') = 0, \quad z_i \varepsilon A_{(l/2)+i}, \quad z_i' \varepsilon A_{(l/2)+i} \quad (1 \leqq i \leqq l/2).$

Equation (18) has $l/2$ summands on the left-hand side, i. e., half as many as the original equation (14).

We set

$$\sum_{i=1}^{l/4}(z_i - z_i') = x, \qquad \sum_{i=(l/4)+1}^{l/2}(z_i - z_i') = y,$$

and thereby bring equation (18) into the form

$$x + y = 0.$$

To this we can apply Lemma 4 anew. It is evident that, just as we arrived at equation (18) from equation (14), we now get from equation (18) to the equation

(19) $$\sum_{i=1}^{l/4}(u_i + u_i' - u_i'' - u_i''') = 0,$$

where we have to consider the sum of the numbers of solutions of this equation under the following (now four) hypotheses:

$$\left.\begin{array}{l} 1)\ u_i, u_i', u_i'', u_i''' \varepsilon A_i, \\ 2)\ u_i, u_i', u_i'', u_i''' \varepsilon A_{(l/4)+i}, \\ 3)\ u_i, u_i', u_i'', u_i''' \varepsilon A_{(l/2)+i}, \\ 4)\ u_i, u_i', u_i'', u_i''' \varepsilon A_{(3l/4)+i}. \end{array}\right\} \quad (1 \leqq i \leqq l/4)$$

Since $l = k \cdot 2^s$, we can repeat this process s times. We evidently end up then with the equation

(20) $$\sum_{i=1}^{k}\{y_i^{(1)} + y_i^{(2)} + \ldots + y_i^{(2^{s-1})} - y_i^{(2^{s-1}+1)} - \ldots - y_i^{(2^s)}\} = 0,$$

where we have to consider the sum of the numbers of solutions of this equation under 2^s different hypotheses, viz.:

$$\left.\begin{array}{l} 1)\ y_1^{(j)}\varepsilon A_1,\ y_2^{(j)}\varepsilon A_2, \dots, y_k^{(j)}\varepsilon A_k, \\ 2)\ y_1^{(j)}\varepsilon A_{k+1},\ y_2^{(j)}\varepsilon A_{k+2}, \dots, y_k^{(j)}\varepsilon A_{2k}, \\ \cdot\ \cdot\ \cdot\ \cdot\ \cdot\ \cdot\ \cdot\ \cdot\ \cdot\ \cdot\ \cdot\ \cdot\ \cdot\ \cdot\ \cdot\ \cdot\ \cdot\ \cdot \\ 2^s)\ y_1^{(j)}\varepsilon A_{k2^s-k+1}, \dots, y_k^{(j)}\varepsilon A_{k2^s}. \end{array}\right\} (1 \leqq j \leqq 2^s)$$

If we put

$$y^{(j)} = y_1^{(j)} + y_2^{(j)} + \dots + y_k^{(j)} \qquad (1 \leqq j \leqq 2^s),$$

then equation (20) takes on the simple form

$$(21) \qquad y^{(1)} + y^{(2)} + \dots + y^{(2^{s-1})} - y^{(2^{s-1}+1)} - \dots - y^{(2^s)} = 0.$$

Here we are concerned with the sum of the numbers of solutions of equation (21) under the following 2^s hypotheses, which differ from one another in the value of the parameter w $(0 \leqq w \leqq 2^s - 1)$:

$$y^{(j)} = y_1^{(j)} + y_2^{(j)} + \dots + y_k^{(j)},$$

where

$$y_1^{(j)}\varepsilon A_{wk+1},\ y_2^{(j)}\varepsilon A_{wk+2},\ \dots,\ y_k^{(j)}\varepsilon A_{(w+1)k} \quad (j = 1, 2, \dots, 2^s).$$

Thus we can express the final result of our deduction in the form of the following proposition:

LEMMA 5. *If $l = k \cdot 2^s$, the number of solutions of the equation*

$$(14) \qquad x_1 + x_2 + \dots + x_l = c, \qquad x_i \varepsilon A_i \quad (1 \leqq i \leqq l)$$

does not exceed the sum of the numbers of solutions of the equation

$$(21) \qquad y^{(1)} + y^{(2)} + \dots + y^{(2^{s-1})} - y^{(2^{s-1}+1)} - \dots - y^{(2^s)} = 0,$$

$$\left.\begin{array}{c} y^{(j)} = y_1^{(j)} + y_2^{(j)} + \dots + y_k^{(j)}, \\ y_1^{(j)}\varepsilon A_{wk+1},\ y_2^{(j)}\varepsilon A_{wk+2},\ \dots,\ y_k^{(j)}\varepsilon A_{(w+1)k} \end{array}\right\} (j = 1, 2, \dots, 2^s)$$

under the hypotheses $w = 0, 1, \dots, 2^s - 1$.

Notice the connection between Lemma 4 and Lemma 5 for $k = s = 1$, $l = 2$.

This winds up our preliminaries, and we are ready now to begin the direct assault on the fundamental lemma.

§5
PROOF OF THE FUNDAMENTAL LEMMA

We are going to prove the fundamental lemma by the method of induction on n. It is often the case in inductive proofs, that a strengthening of the proposition to be proved, considerably facilitates its proof by the given method (and sometimes is actually what makes the proof feasible in the first place). The reason for this is easy to understand. In inductive proofs, the proposition is assumed to be correct for the number $n-1$, and is proved for the number n. Hence, the stronger the proposition, the more that is given to us by the case $n-1$; of course, so much the more has to be proved for the number n, but in many problems the first consideration turns out to be more important than the second.

And so it is, in fact, in the present case. Of immediate interest for us is the number of solutions of the equation $x_1^n+x_2^n+\ldots+x_k^n=m$ $(1\leqq m\leqq N)$ (where, according to the very meaning of the problem, $0\leqq x_i\leqq m^{1/n}\leqq N^{1/n}$). But x^n is the simplest special case of an n-th degree polynomial

$$f(x)=a_0x^n+a_1x^{n-1}+\ldots+a_{n-1}x+a_n,$$

and it will be to our advantage to replace the given equation (1) by the more general equation

$$f(x_1)+f(x_2)+\ldots+f(x_k)=m, \tag{22}$$

where the unknowns are subjected to the weaker conditions $|x_i|\leqq N^{1/n}$ $(1\leqq i\leqq k)$. The proof of our proposition for equation (22) will give us more than we really need; but, as you will see, it is just this strengthening of our proposition which creates the possibility of an induction. And so, for $m\leqq N$, let us denote by $r_k(m)$ the number of solutions of equation (22) which satisfy the conditions $|x_i|\leqq N^{1/n}$ $(1\leqq i\leqq k)$. Of course we are still free to dispose arbitrarily of the coefficients of the polynomial $f(x)$ in the interest of the induction to be performed (provided only that the imposed conditions are satisfied in the case $f(x)=x^n$). We are going to prove the following proposition:

Let the coefficients of the polynomial $f(x)$ satisfy the inequalities

$$|a_i|\leqq c(n)N^{i/n} \qquad (0\leqq i\leqq n). \tag{23}$$

Then, for a suitably chosen $k=k(n)$,

$$r_k(m) < c(n)\,N^{(k/n)-1} \quad (1 \leq m \leq N).$$

Since the inequalities (23) are obviously satisfied in the case $f(x)=x^n$ for $c(n)=1$, this theorem is indeed a sharpening of our fundamental lemma.

Let us first consider the case $n=1$, $f(x)=a_0x+a_1$. We set $k(1)=2$, so that equation (22) acquires the form

$$a_0(x_1+x_2)=m-2a_1.$$

We are interested in solutions of this equation which satisfy the requirements $|x_1| \leq N$, $|x_2| \leq N$. Thus at most $2N+1 \leq 3N$ values are possible for x_1. But at most one x_2 corresponds to every x_1, so that

$$r_2(m) \leq 3N,$$

which completes the proof of our proposition for $n=1$ $(k=2)$.

Now let $n>1$, and suppose that our assertion has already been verified for the exponent $n-1$. Put $k(n-1)=k'$ and choose

$$k=k(n)=2n\cdot 2^{[4\log_2 k']},$$

where the exponent means the greatest integer not exceeding $4\log_2 k'$. In the sequel we shall set $[4\log_2 k']-1=s$, for brevity, so that

(24) $$k=2n\cdot 2^{s+1}.$$

To estimate the number, $r_k(m)$, of solutions of equation (22), we first apply Lemma 4 to it, setting

$$x=\sum_{i=1}^{\frac{1}{2}k} f(x_i), \quad y=\sum_{i=\frac{1}{2}k+1}^{k} f(x_i).$$

The complex A (and the complex B which coincides with it in this case) consists of all sums of the form

$$\sum_{i=1}^{\frac{1}{2}k} f(x_i), \text{ where } |x_i| \leq N^{1/n} \ (1 \leq i \leq \tfrac{1}{2}k).$$

By the Corollary of Lemma 4, $r_k(m)$ does not exceed the number of solutions of the equation $x-y=0$, where $x \in A$, $y \in A$, i. e.,

$$x=\sum_{i=1}^{k/2} f(x_i), \quad y=\sum_{i=1}^{k/2} f(y_i),$$

$$|x_i| \leqq N^{1/n}, \quad |y_i| \leqq N^{1/n} \quad (1 \leqq i \leqq k/2).$$

In other words, $r_k(m)$ does not exceed the number of solutions of the equation

$$\sum_{i=1}^{k/2} \{f(x_i)-f(y_i)\}=0, \tag{25}$$

where $|x_i| \leqq N^{1/n}$, $|y_i| \leqq N^{1/n}$ $(1 \leqq i \leqq k/2)$. We now set $x_i - y_i = h_i$ $(1 \leqq i \leqq k/2)$ and replace the system of unknowns $\{x_i, y_i\}$ by the system $\{y_i, h_i\}$; here we allow y_i and h_i $(1 \leqq i \leqq k/2)$ to assume all possible integral values in the interval $<-2N^{1/n}, +2N^{1/n}>$, which can only increase the number of solutions of our equation. This means that every summand $f(x_i)-f(y_i)$ in equation (25) is replaced by the expression

$$\begin{aligned} f(y_i+h_i)-f(y_i) &= \sum_{v=0}^{n-1} a_v \{(y_i+h_i)^{n-v} - y_i^{n-v}\} \\ &= \sum_{v=0}^{n-1} a_v \sum_{t=1}^{n-v} \binom{n-v}{t} h_i^t y_i^{n-v-t}. \end{aligned}$$

If we change the variable t of summation by putting

$$v+t=u,$$

so that

$$n-v-t=n-u, \qquad t=u-v,$$

we obtain

$$\begin{aligned} f(y_i+h_i)-f(y_i) &= h_i \sum_{v=0}^{n-1} a_v \sum_{u=v+1}^{n} \binom{n-v}{u-v} h_i^{u-v-1} y_i^{n-u} \\ &= h_i \sum_{u=1}^{n} y_i^{n-u} \sum_{v=0}^{u-1} a_v \binom{n-v}{u-v} h_i^{u-v-1} \\ &= h_i \sum_{u=1}^{n} a_{i,u} y_i^{n-u} = h_i \phi_i(y_i), \end{aligned}$$

where

$$\phi_i(y) = \sum_{u=1}^{n} a_{i,u} y^{n-u}$$

is a polynomial of degree $n-1$ with coefficients

$$a_{i,u}=\sum_{v=0}^{u-1} a_v \binom{n-v}{u-v} h_i^{u-v-1} \quad (1 \leqq i \leqq k/2)$$

which depend on the numbers h_i.

Thus, in the new variables $\{y_i, h_i\}$, equation (25) assumes the form

$$h_1\phi_1(y_1)+h_2\phi_2(y_2)+\ldots+h_{\frac{1}{2}k}\phi_{\frac{1}{2}k}(y_{\frac{1}{2}k})=0. \tag{26}$$

In this equation the numbers h_i and y_i may take on arbitrary integral values in the interval $\langle -2N^{1/n}, +2N^{1/n}\rangle$, where we must bear in mind that the coefficients of the polynomials $\phi_i(y)$ (of degree $n-1$) depend on the numbers h.

Mark well that we have proved the following so far: *The number $r_k(m)$ which we are estimating, does not exceed the sum of the numbers of solutions in integers y_i, $|y_i| \leqq 2N^{1/n}$ $(1 \leqq i \leqq k/2)$, of all the equations* (26) *which can be obtained from all possible values of the numbers h_i, $|h_i| \leqq 2N^{1/n}$ $(1 \leqq i \leqq k/2)$.*

§ 6

CONTINUATION

We are now going to examine one of the equations (26), i.e., we shall regard the numbers h_i $(1 \leqq i \leqq k/2)$ for a while as fixed. Let us apply Lemma 5 to this equation; the numbers $h_i\phi_i(y_i)$ play the role of the unknowns x_i, the number $\frac{1}{2}k = 2n \cdot 2^s$ plays the role of the number l, and we set $2n = k_0$ for brevity. Recall once more that the numbers h_i appear in equation (26) not only explicitly but also through the coefficients of the polynomials $\phi_i(y)$. The complex A_i to which the numbers $x_i = h_i\phi_i(y_i)$ must belong consists, in the present case, of all numbers of the form $h_i\phi_i(y_i)$, where the numbers h_i have given, fixed values and the numbers y_i run through the interval $\langle -2N^{1/n}, +2N^{1/n}\rangle$.

According to Lemma 5, the number of solutions of equation (26) satisfying the requirements just described, does not exceed the sum of the numbers of solutions of the equation

$$y^{(1)}+y^{(2)}+\ldots+y^{(2^{s-1})}-y^{(2^{s-1}+1)}-\ldots-y^{(2^s)}=0 \tag{21}$$

under the following 2^s hypotheses which correspond to the values of the parameter $w=0, 1, \ldots, 2^s-1$:

$$\left.\begin{array}{l} y^{(j)}=y_1^{(j)}+y_2^{(j)}+\ldots+y_{k_0}^{(j)}, \\ y_i^{(j)}\in A_{wk_0+i} \quad (1\leqq i\leqq k_0) \end{array}\right\} \qquad (1\leqq j\leqq 2^s),$$

where, remember, A_r $(1\leqq r\leqq 2^s)$ is the complex of numbers of the form $h_r\phi_r(y_r)$ with prescribed h_r and arbitrary y_r, $|y_r|\leqq 2N^{1/n}$.

For the case $w=0$ (which we choose merely as an example), equation (21) in expanded form looks as follows:

$$\begin{aligned} &\{y_1^{(1)}+y_2^{(1)}+\ldots+y_{k_0}^{(1)}\} \\ &+\{y_1^{(2)}+y_2^{(2)}+\ldots+y_{k_0}^{(2)}\} \\ &+\cdot\;\cdot\;\cdot\;\cdot\;\cdot\;\cdot\;\cdot\;\cdot\;\cdot \\ &+\{y_1^{(2^{s-1})}+y_2^{(2^{s-1})}+\ldots+y_{k_0}^{(2^{s-1})}\} \\ &-\{y_1^{(2^{s-1}+1)}+y_2^{(2^{s-1}+1)}+\ldots+y_{k_0}^{(2^{s-1}+1)}\} \\ &-\cdot\;\cdot\;\cdot\;\cdot\;\cdot\;\cdot\;\cdot\;\cdot\;\cdot\;\cdot\;\cdot\;\cdot \\ &-\{y_1^{(2^s)}+y_2^{(2^s)}+\ldots+y_{k_0}^{(2^s)}\}=0, \end{aligned}$$

or, rearranging the summands,

$$\begin{aligned} &\{y_1^{(1)}+y_1^{(2)}+\ldots+y_1^{(2^{s-1})}-y_1^{(2^{s-1}+1)}-\ldots-y_1^{(2^s)}\} \\ &+\{y_2^{(1)}+y_2^{(2)}+\ldots+y_2^{(2^{s-1})}-y_2^{(2^{s-1}+1)}-\ldots-y_2^{(2^s)}\} \\ &+\cdot\;\cdot\;\cdot\;\cdot\;\cdot\;\cdot\;\cdot\;\cdot\;\cdot\;\cdot\;\cdot\;\cdot\;\cdot\;\cdot \\ &+\{y_{k_0}^{(1)}+y_{k_0}^{(2)}+\ldots+y_{k_0}^{(2^{s-1})}-y_{k_0}^{(2^{s-1}+1)}-\ldots-y_{k_0}^{(2^s)}\}=0; \end{aligned}$$

every one of the numbers $y_i^{(j)}$ is a number of the form $h_i\phi_i(v_i^{(j)})$, where $|v_i^{(j)}|\leqq 2N^{1/n}$. Hence the last equation can be rewritten in the form

$$\begin{aligned} &h_1\{\phi_1(v_1^{(1)})+\phi_1(v_1^{(2)})+\ldots+\phi_1(v_1^{(2^{s-1})})-\phi_1(v_1^{(2^{s-1}+1)})-\ldots-\phi_1(v_1^{(2^s)})\} \\ &+h_2\{\phi_2(v_2^{(1)})+\ldots-\phi_2(v_2^{(2^s)})\}+\ldots+h_{k_0}\{\phi_{k_0}(v_{k_0}^{(1)})+\ldots-\phi_{k_0}(v_{k_0}^{(2^s)})\}=0. \end{aligned}$$

By putting, for brevity,

$$\phi_i(v_i^{(1)})+\phi_i(v_i^{(2)})+\ldots+\phi_i(v_i^{(2^{s-1})})-\phi_i(v_i^{(2^{s-1}+1)})-\ldots-\phi_i(v_i^{(2^s)})=z_i, \quad (1\leqq i\leqq k_0)$$

this equation can be written quite shortly as follows:

$$h_1z_1+h_2z_2+\ldots+h_{k_0}z_{k_0}=0. \tag{27}$$

All together we have 2^s equations of this sort, and their totality can be written down in the compact form

$$\sum_{i=1}^{k_0} h_{w\,k_0+i}\,z_{w\,k_0+i}=0 \qquad (0\leqq w\leqq 2^s-1).$$

For the present, however, we shall confine our investigation to equation (27), which may of course be regarded as typical. To estimate the number of solutions of this equation which interest us, we must first see within what limits the quantity $\phi_i(v_i^{(j)})$ can vary. To this end we recall that (p. 55)

$$\phi_i(y)=\sum_{u=1}^{n} a_{i,u}\,y^{n-u},$$

where

$$a_{i,u}=\sum_{v=0}^{u-1} a_v\binom{n-v}{u-v}h_i^{u-v-1} \qquad (1\leqq i\leqq k/2).$$

Hence it follows from our hypotheses $|a_v|<c(n)N^{v/n}$ and $|h_i|\leqq 2N^{\frac{1}{n}}$, that

$$|a_{i,u}|<\sum_{v=0}^{u-1} c(n)N^{v/n}\binom{n-v}{u-v}c(n)N^{(u-v-1)/n}=c(n)N^{(u-1)/n}\sum_{v=0}^{u-1}\binom{n-v}{u-v};$$

i. e., in view of $u\leqq n$,

$$|a_{i,u}|<c(n)\,N^{(u-1)/n}. \tag{28}$$

On the other hand, because of $|v_i^{(j)}|\leqq 2N^{1/n}$, we have $|v_i^{(j)}|^{n-u}\leqq c(n)\cdot N^{(n-u)/n}$ and consequently

$$|a_{i,u}|\cdot|v_i^{(j)}|^{n-u}\leqq c(n)\,N^{(u-1)/n}\,N^{(n-u)/n}=c(n)N^{(n-1)/n}.$$

The same estimate (with another $c(n)$) holds for the whole $\phi_i(v_i^{(j)})$, since the number of terms of this polynomial is equal to n. Accordingly

$$|\phi_i(v_i^{(j)})|<c(n)\,N^{(n-1)/n} \quad (1\leqq i\leqq k_0, 1\leqq j\leqq 2^s).$$

But every z_i is the sum of $2^s = c(n)$ summands of the form $\pm\phi_i(v_i^{(j)})$, and therefore

$$|z_i| < c(n)N^{(n-1)/n} \qquad (1 \leq i \leq k_0)$$

(with another $c(n)$, naturally). This means that in equation (27) every z_i can assume only the values lying in the interval $\langle -c(n)N^{(n-1)/n}, +c(n)N^{(n-1)/n}\rangle$.

Let $\bar{m}$ be one of these numbers. The equation $z_i = \bar{m}$ can be satisfied in general not only in one but in several ways, because the definition of the number z_i (p. 58) is such that one and the same value of z_i can very well result from different choices of the numbers $v_i^{(j)}$ $(1 \leq j \leq 2^s)$. We now have to estimate the number of solutions of the relation $z_i = \bar{m}$, i. e., of the equation

$$(29) \quad \phi_i(v_i^{(1)}) + \ldots + \phi_i(v_i^{(2^{s-1})}) - \phi_i(v_i^{(2^{s-1}+1)}) - \ldots - \phi_i(v_i^{(2^s)}) = \bar{m}.$$

For this purpose we shall finally have to apply the long-promised induction. We proceed as follows.

First we rewrite equation (29) in the form

$$\phi_i(v_i^{(1)}) + \phi_i(v_i^{(2)}) + \ldots + \phi_i(v_i^{(k')})$$
$$= \bar{m} - \phi_i(v_i^{(k'+1)}) - \ldots + \phi_i(v_i^{(2^{s-1}+1)}) + \ldots + \phi_i(v_i^{(2^s)}).$$

This is possible because for $k' = k(n-1) > 1$ (and we have seen that already $k(1) = 2$) we have

$$2^{s-1} = 2^{[4 \log_2 k'] - 2} > k'.$$

(In detail: $k' \geq 2$, $\log_2 k' \geq 1$, $3 \log_2 k' \geq 3$, $[4 \log_2 k'] - 2 > 4 \log_2 k' - 3 \geq \log_2 k'$, $2^{s-1} = 2^{[4 \log_2 k'] - 2} \geq k'$.)

If we denote the right-hand side of the last equation by m', we get

$$(30) \qquad \phi_i(v_i^{(1)}) + \ldots + \phi_i(v_i^{(k')}) = m'.$$

Let us choose some particular values for the numbers

$$v_i^{(j)} \qquad (k'+1 \leq j \leq 2^s)$$

(in the interval $\langle -2N^{1/n}, +2N^{1/n}\rangle$, naturally); then m' also acquires a definite value. To equation (30) we now apply the theorem to be proved, since $\phi_i(y)$ is a polynomial of degree $n-1$. We have to verify that all the necessary hypotheses are fulfilled. We have

$$\phi_i(y) = \sum_{u=1}^{n} a_{i,u} y^{n-u},$$

where, according to (28),

$$|a_{i,u}| < c(n) N^{\frac{u-1}{n}} = c(n)(N^{\frac{n-1}{n}})^{\frac{u-1}{n-1}}, \tag{31}$$

and, as is easily seen,

$$|m'| < c(n) N^{\frac{n-1}{n}}$$

(because $\bar{m}$ and all $\phi_i(y_i^{(j)})$ satisfy this inequality).

In virtue of the last inequality, the role of N can be assumed by the number $c(n)N^{(n-1)/n}$; then the conditions (31), which the coefficients of the polynomial $\phi_i(y)$ satisfy, are precisely the conditions (23) with n replaced by $n-1$. Thus all the hypotheses are indeed fulfilled, and we can assert that the number of solutions of equation (30), for which $|v_i^{(j)}| \leqq 2N^{1/n} = 2(N^{(n-1)/n})^{1/(n-1)}$, does not exceed the number

$$c(n)(N^{\frac{n-1}{n}})^{\frac{k'}{n-1}-1} = c(n) N^{\frac{k'-n+1}{n}}. \tag{32}$$

This estimate is obtained for the fixed values $v_i^{(k'+1)}, \ldots, v_i^{(2^s)}$. Clearly we have at most

$$(4N^{1/n}+1)^{2^s-k'} < c(n) N^{\frac{2^s-k'}{n}} \tag{33}$$

such systems of values. The total number of solutions of the required type, of equation (29), therefore does not exceed the product of the right-hand sides of (32) and (33), i. e., it is at most

$$c(n) N^{\frac{2^s-n+1}{n}}. \tag{34}$$

We now return to equation (27). We saw before (p. 59) that every z_i can assume only the values lying in the interval $\langle -c(n)N^{(n-1)/n}, +c(n)N^{(n-1)/n}\rangle$. Now we see that the "multiplicity" of each of these values (i. e., the number of ways of choosing the $y_i^{(j)}$ so as to satisfy the equation) does not exceed the number (34).

This result makes it possible to reduce the whole problem to an estimation of the numbers of solutions of linear equations. For at the end of §5 we reduced the estimation of $r_k(m)$ to the estimation of the numbers of solutions of equations of the form (26). But as we proved by an application of Lemma 5, the number of solutions of equation (26), for which $|y_i| \leqq 2N^{1/n}$, is at most equal to the sum of the numbers of solutions of 2^s equations of type (27), i. e., already linear equations. In this connection we obtained limits within which the unknowns z_i are allowed to vary. A certain new difficulty (the price we have had to pay for the transition to linear equations) is that the new unknowns z_i have to be considered with certain multiplicities (for which we have also determined limits).

Finally we must not forget that all these calculations are made under the assumption that the numbers h_i are chosen and fixed. Therefore we still have to multiply the result obtained, by the number of all such possible choices.

The final result of this section, which we have to keep in mind, reads: *Our estimated number $r_k(m)$ does not exceed the sum of the numbers of solutions in integers z_i, $|z_i| \leqq c(n)N^{(n-1)/n}$, with multiplicities $\lambda_i \leqq c(n)N^{(2^s-n+1)/n}$, of equations of the form*

$$\sum_{i=1}^{k_0} h_{wk_0+i}\, z_{wk_0+i} = 0, \tag{35}$$

where w runs through the values $0, 1, \ldots, 2^s-1$, and the numbers h_r $(1 \leqq r \leqq 2^s k_0)$ assume, independently of one another, all integers in the interval $<-2N^{1/n}, +2N^{1/n}>$.

And so we see that we have now obtained an estimate for $r_k(m)$, in whose formulation the given polynomial $f(x)$ does not appear, which lends this estimate a very general character.

§7
CONCLUSION

Now that we have reduced the problem to an estimation of the number of solutions of linear equations which are independent of the special form of the polynomial $f(x)$, we quickly reach our goal with the aid of Lemma 3.

Denote by K any particular combination of the numbers h_i,

$|h_i| \leq 2N^{1/n}$ $(1 \leq i \leq k/2)$, and by $U_w(\mathrm{K})$ the number of solutions of equation (35) for this fixed combination K and for a certain prescribed w, where we are concerned with those solutions z_i which satisfy the inequalities $|z_i| \leq c(n)N^{(n-1)/n}$, with multiplicities $\lambda_i \leq c(n)N^{(2^s-n+1)/n}$. Then, according to the final result in the preceding section,

$$r_k(m) \leq \sum_{\mathrm{K}} \{ \sum_{w=0}^{2^s-1} U_w(\mathrm{K}) \},$$

where the summation over K extends over all admissible combinations K of the numbers h_i. This can be written

$$r_k(m) \leq \sum_{w=0}^{2^s-1} \{ \sum_{\mathrm{K}} U_w(\mathrm{K}) \}.$$

It is immediately evident, however, that for different w the sums $\sum_{\mathrm{K}} U_w(\mathrm{K})$ do not differ from one another at all (because for different w the equations (35) do not differ from one another in any respect). We can therefore write

$$r_k(m) \leq 2^s \sum_{\mathrm{K}} U_0(\mathrm{K}) = c(n) \sum_{\mathrm{K}} U_0(\mathrm{K}).$$

Here $U_0(\mathrm{K})$ is the number of solutions of the equation

(36) $$h_1 z_1 + h_2 z_2 + \ldots + h_{k_0} z_{k_0} = 0$$

for the given combination K of the numbers h_i, $|h_i| \leq 2N^{1/n}$ $(1 \leq i \leq k/2)$, where $|z_i| \leq c(n)N^{(n-1)/n}$ and the z_i have multiplicities $\lambda_i \leq c(n) \cdot N^{(2^s-n+1)/n}$. Let us denote by $U_0^*(\mathrm{K})$ the number of solutions of the same equation under the assumption that all z_i are simple. Then clearly

$$U_0(\mathrm{K}) \leq \{ c(n) N^{\frac{2^s-n+1}{n}} \}^{k_0} U_0^*(\mathrm{K}),$$

or, recalling that $k_0 = 2n$,

$$U_0(\mathrm{K}) \leq c(n) N^{2(2^s-n+1)} U_0^*(\mathrm{K}),$$

and hence

(37) $$r_k(m) \leq c(n) N^{2(2^s-n+1)} \sum_{\mathrm{K}} U_0^*(\mathrm{K}).$$

Now let us note the following. Every K represents a certain admissible combination of the values of all h_i $(1 \leq i \leq k/2)$; the number $U_0^*(\mathrm{K})$, however, is completely determined by the values of the first $k_0 = 2n$ of these values $(1 \leq i \leq 2n)$, because they alone appear in equation (36). Of course when we choose a certain fixed combination K, we thereby also uniquely define a certain combination K′ of the values $h_1, h_2, \ldots, h_{2n}$. But if, conversely, a certain combination K′ of the numbers $h_1, h_2, \ldots, h_{2n}$ is selected, there corresponds to it not the single combination K, but rather as many as there are ways of choosing the remaining "supplements" h_i $(2n < i \leq k/2)$. Since every h_i must belong to the interval $\langle -2N^{1/n}, +2N^{1/n} \rangle$, it is evident that to a combination K′ there correspond at most

$$c(n)\,(N^{1/n})^{(k/2)-2n} = c(n)\,N^{(k/2n)-2}$$

combinations K. Hence

$$\sum_{\mathrm{K}} U_0^*(\mathrm{K}) \leq c(n)\,N^{(k/2n)-2} \sum_{\mathrm{K}'} U_0^*(\mathrm{K}'),$$

where $U_0^*(\mathrm{K}')$ is the number of solutions in integers z_i, $|z_i| \leq c(n) \cdot N^{(n-1)/n}$ $(1 \leq i \leq 2n)$ of equation (36) for the given combination K′ of the numbers h_i, $|h_i| \leq 2N^{1/n}$ $(1 \leq i \leq 2n)$, and the summation is to be extended over all such combinations. From (37) we therefore obtain*

$$(38)\quad r_k(m) \leq c(n) N^{2(2^s-n+1)} N^{(k/2n)-2} \sum_{\mathrm{K}'} U_0^*(\mathrm{K}') = c(n) N^{2(2^{s+1}-n)} \sum_{\mathrm{K}'} U_0^*(\mathrm{K}').$$

Finally, $\sum_{\mathrm{K}'} U^*(\mathrm{K}')$ is immediately estimated with the help of Lemma 3, where we have to put $l = 2n$, $A = 2N^{1/n}$, $B = c(n) N^{(n-1)/n}$; you can easily verify that all the hypotheses of Lemma 3 are satisfied. On applying this lemma we find that

$$\sum_{\mathrm{K}'} U_0^*(\mathrm{K}') \leq c(n)\,(AB)^{2n-1} = c(n) N^{2n-1}.$$

At last, inequality (38) yields

$$r_k(m) \leq c(n) N^{2(2^{s+1}-n)} \cdot N^{2n-1} = c(n) N^{2 \cdot 2^{s+1}-1} = c(n) N^{\frac{k}{n}-1},$$

which completes the proof of the fundamental lemma and thereby also of Hilbert's theorem.

This proof, so exquisitely elementary, will undoubtedly seem very complicated to you. But it will take you only two to three

*Recall that $k = 2n \cdot 2^{s+1}$.

weeks' work with pencil and paper to understand and digest it completely. It is by conquering difficulties of just this sort, that the mathematician grows and develops.

* * *